KU-752-820

Contents

Acknowledgements

The author and publishers wish to thank the following who have kindly given permission for the use of copyright material.

The Associated Examining Board, East Anglian Examinations Board, East Midland Regional Examinations Board, Northern Ireland Schools Examinations Council, University of Cambridge Local Examinations Syndicate, University of London School Examinations Board, University of Oxford Delegacy of Local Examinations, The Southern Regional Examinations Board, The London Regional Examining Board, The West Midlands Examinations Board, and Yorkshire and Humberside Regional Examinations Board for questions from past examination papers.

The University of Cambridge Local Examinations Syndicate bears no responsibility for the answers to questions taken from its past question-papers which are contained in this publication.

Every effort has been made to trace all the copyright holders but if any have been inadvertently overlooked the publishers will be pleased to make the necessary arrangement at the first opportunity.

The University of London Entrance and School Examinations Council accepts no responsibility whatsoever for the accuracy or method in the answers given in this book to actual questions set by the London Board.

Acknowledgement is made to the Southern Universities' Joint Board for School Examinations for permission to use questions taken from their past papers but the Board is in no way responsible for answers that may be provided and they are solely the responsibility of the authors.

The Associated Examining Board, the University of Oxford Delegacy of Local Examinations, the Northern Ireland Schools Examination Council and the Scottish Examination Board wish to point out that worked examples included in the text are entirely the responsibility of the author and have neither been provided nor approved by the Board.

I am grateful for the permission of previous collaborators for the use of certain illustrations and text from Macmillan books, namely *Biology: An Integrated Approach* (with S. T. Smith) and *Certificate Model Answers (Biology)* (with N. Green).

I am also grateful for the initiative and assistance of Mary Waltham, Senior Editor of Macmillan Education, and finally to my wife for her help, patience and forbearance over many years of writing.

Cover photograph courtesy of S and G Photograph Library.

Organisations Responsible for GCSE Examinations

In the United Kingdom, examinations are administered by the following organisations. Syllabuses and examination papers can be ordered from the addresses given here:

Northern Examining Association (NEA)

Joint Matriculation Board (JMB)
　Publications available from:
John Sherratt & Son Ltd
78 Park Road, Altrincham
Cheshire WA14 5QQ

North Regional Examinations Board
Wheatfield Road, Westerhope
Newcastle upon Tyne NE5 5JZ

Yorkshire and Humberside Regional
　Examinations Board (YREB)
Scarsdale House, 136 Derbyside Lane,
Sheffield S8 8SE

Associated Lancashire Schools
　Examining Board
12 Harter Street
Manchester M1 6HL

North West Regional
　Examinations Board (NWREB)
Orbit House, Albert Street
Eccles, Manchester M30 0WL

Midland Examining Group (MEG)

University of Cambridge Local
　Examinations Syndicate (UCLES)
Syndicate Buildings, Hills Road
Cambridge CB1 2EU

Oxford and Cambridge Schools
　Examination Board (O & C)
10 Trumpington Street
Cambridge CB2 1QB

Southern Universities' Joint Board
　(SUJB)
Cotham Road
Bristol BS6 6DD

East Midland Regional Examinations
　Board (EMREB)
Robins Wood House, Robins Wood Road
Aspley, Nottingham NG8 3NR

West Midlands Examinations Board
　(WMEB)
Norfolk House, Smallbrook
Queensway, Birmingham B5 4NJ

London and East Anglian Group (LEAG)

University of London School
　Examinations Board (L)
University of London Publications Office
52 Gordon Square, London WC1E 6EE

London Regional Examining Board
　(LREB)
Lyon House
104 Wandsworth High Street
London SW18 4LF

East Anglian Examinations Board (EAEB)
The Lindens, Lexden Road
Colchester, Essex CO3 3RL

Southern Examining Group (SEG)

The Associated Examining Board (AEB)
Stag Hill House
Guildford, Surrey GU2 5XJ

**University of Oxford Delegacy of
  Local Examinations (OLE)**
Ewert Place, Banbury Road
Summertown, Oxford OX2 7BZ

**Southern Regional Examinations
  Board (SREB)**
Avondale House, 33 Carlton Crescent
Southampton, Hants SO9 4YL

**South-East Regional Examinations
  Board (SEREB)**
Beloe House
2–10 Mount Ephraim Road
Royal Tunbridge Wells, Kent TN1 1EU

**South-Western Examinations Board
  (SWExB)**
23–29 Marsh Street, Bristol BS1 4BP

Scottish Examination Board (SEB)
  Publications available from:
Robert Gibson and Sons (Glasgow) Ltd
17 Fitzroy Place, Glasgow G3 7SF

Welsh Joint Education Committee (WJEC)
245 Western Avenue
Cardiff CF5 2YX

Northern Ireland Schools Examinations Council (NISEC)
Examinations Office, Beechill House
Beechill Road, Belfast BT8 4RS

1 Introduction

1.1 How to use this book

Each chapter provides the following:

1. A short summary of the main facts, terms and principles associated with each topic area. This includes

    (a) key words in **bold**;
    (b) definitions;
    (c) summary tables, comparative tables and diagrams.

2. A selection of past examination questions:

    (a) multiple-choice questions – simple completion;
    (b) structured – short answer;
    (c) structured free-response – short answer and essay.

3. Answers to multiple-choice questions and some structured questions.

The summary cannot provide the complete information that would be available in a standard textbook*, and therefore it is suggested that you use this book alongside a textbook of human and social biology. This book should be used throughout a GCSE course, to help in preparing answers for home work and when revising for internal or external examinations. It will help you to familiarise yourself with standard examination questions and to develop practice in examination techniques.

The new GCSE examination is designed to examine and grade across the full ability range of the old CSE and GCE examinations. Questions have therefore been taken, where appropriate, from examination papers of CSE and GCE boards. The GCSE examinations will have differentiated papers, usually involving a common set of papers for all candidates, with additional paper(s) for attaining the higher grades. The questions chosen in each chapter show a range of difficulty, and where relevant the examination board is indicated beside the question.

This book only attempts to assist you with the theoretical examination papers. It does not attempt to include information or examples about practical work and its assessment, with the exception of some questions on practical work which can be answered from theoretical knowledge. The skills and processes of practical biology will be assessed by your teachers and will be a progressive accumulation of marks as you proceed through your course of practical work. At the end of this introductory chapter is listed some of the experimental work which may be used to assess your skill and understanding (see Section 1.8).

The purpose of providing answers is to give a guide to the length and breadth of knowledge required for the question posed and to give you an idea of what the examiner is looking for in a particular question. In some cases the structured questions have the answers entered directly on the question, but, in general, multiple-choice and structured questions have the answers separately recorded at

*For example, *Modern Human and Social Biology*, Soper and Smith, Macmillan.

the end of the Questions and Answers section. Multiple-choice objective questions have only one correct response, but structured questions often have a number of answers that would be acceptable. This applies even to 'one-word' answers and, of course, is particularly applicable to 'essay-type' answers, where candidates may write their answer in a variety of ways.

1.2 Initial preparation for the examination

To be of any real value, revision must be effective. For this to be so, as well as having an efficient revision technique, the candidate should be fit and fresh. This state can only be achieved by taking regular exercise and having plenty of sleep.

(a) The Syllabus

It is useful to obtain a syllabus and use it to plan a revision timetable. Ensure that the timetable covers **all aspects** of the syllabus. Allow eight to twelve weeks for **complete revision** of the subject.

(b) Revision

The candidate will get the best out of him/herself if he/she works **without distraction**, for periods of one hour or so. After this time, relax, go for a walk, play a game, etc., and then return refreshed for a further period of study.

All sensory channels should be used for revision.

1. **Speak the material**, either to yourself, to friends who are studying the same subject or into a tape recorder, should one be available, for later playback and comparison with the text.
2. **Write out the material** and practise drawing. Always compare these efforts with the information given in the texts and continue to practise until the work becomes accurate.
3. **Use memory aids** such as key words or phrases where appropriate.
4. Obtain **copies of previous examination papers**. Notice the style in which the questions are set. Work through the questions and then look up the parts which you have answered incorrectly or have not remembered.
5. It is often useful to put **revision notes on index cards**. These can be used for final revision just before the examination. Making individual notes is certainly preferable to relying on ready-prepared ones!

ABOVE ALL, ATTEMPT TO UNDERSTAND ALL WORK THAT IS BEING LEARNED.

1.3 The examination day

1. Ensure that you know the date, time and place for each paper of the examination. Marks lost from a missed examination can **never** be recovered.
2. Arrive for the examination in plenty of time. It would be most unfortunate for any candidate to be late for an occasion for which he/she has been preparing for the last two or three years.

3. Check that the materials (pen, pencil, ruler, rubber, compass, etc.) required for the examination are all present and correct, and in working order, **before** entering the examination room.
4. When the question paper is given out, fill in the particulars required at the head of the paper, e.g. name, examination number, centre number, number of answer sheet, number of question, etc.
5. Read carefully all instructions. From this the following will be established:

   (a) The time allowed for the whole paper.
   (b) The number of questions that must be answered; **do not answer** more than are required.
   (c) Whether or not there is a compulsory question to be attempted. **Before** the examination check this point by studying previous papers and the syllabus instructions. Always abide by the instructions.

6. Subtract five minutes from the time allowed. This should be used for reading through the question paper. A further five minutes should be used at the end of the examination to check what has been written. Divide the remainder of the time by the number of questions that have to be attempted, to give a rough idea of the length of time which should be spent on each question.
7. At the end of the examination, ensure that the papers of the script are in the correct order and that all of your work is collected.

1.4 Terms used in examinations

Questions normally begin with an instruction designed to guide the candidate to provide a correct response. These guiding terms should be considered carefully and, in order to clarify the meaning, many of them are listed below (in alphabetical order).

1. **Annotated diagrams** – diagrams are required with brief notes alongside. The notes are more than a label. A long written account is unnecessary.
2. **Compare and contrast** – points of similarity between items should be discussed side by side as they arise. Points of contrast could then follow in the same manner.
3. **Define** – a conventional statement or strict definition is all that is required here.
4. **Describe** – putting into words what can actually be observed, or what is understood.
5. **Discuss** – this type of question needs to be planned, as it is easy for the candidate to digress away from the point if due care is not taken. A balanced argument is required. This should be your opinion discussed in a reasonable way.
6. **Distinguish between** – highlight the differences between the materials or processes being discussed.
7. **Give an account of** – a description of a particular process, organ, experiment, etc., is required. Ensure that the particular topic is fully covered, maintaining relevance at all times.
8. **Give an accurate interpretation of** – experimental results must be related to particular processes: for example, respiration.
9. **Give an illustrated account of** – diagrams are essential for this question and also a written account, which must refer to the diagrams presented. Written work and diagrams must complement each other.
10. **Measure** – quantities which can be measured directly from a measuring instrument.

11. **Outline** — the written prose requires only the most important points. A detailed account is not required.
12. **State** — a precise answer is required.
13. **State and explain** — a concise answer amplified with sound reasoning.
14. **Suggest** — the candidate is asked to apply general knowledge to the situation presented. The situation might be novel and the material used by the examiners might not necessarily be interpreted as an integral part of the syllabus. This is where principles have to be applied to unfamiliar situations.
15. **Tabulate** — production of a table showing the main facts required.
16. **What do you understand by** — a definition with the addition of some comment about the significance or relevance of the topic concerned.

1.5 Types of questions

(a) Multiple-choice Questions

There are several types of multiple-choice questions, but most examination boards only set the simple completion type illustrated in this book. The correct answer must be selected by carefully reading each response **A** to **E**. Then the correct response (key response) is entered on a form, or a box is crossed through, for that question number. If the correct answer is not at first apparent, then try to eliminate the four incorrect answers according to the question (the stem). For example:

Some fat is essential in the diet of people living in the tropics because
A fat is a very good source of heat energy.
B some vitamins are soluble only in fat.
C fat forms an insulating layer in the skin.
D the liver can manufacture sugar from fat.
E fat is essential for life in cold climates.

A, C and **E** are wrong and can be eliminated because they refer to retaining heat rather than losing heat, which is the point made specifically in the stem, i.e. 'in the tropics'. **D** is incorrect, so that **B** is the correct response.

(b) Structured Short-answer Questions

The answers to questions of this type are limited, since they must usually be written within the space provided. Thus, a single-word answer may have one line on which to write, while a slightly more extended answer may have two or three lines. Conciseness is most important, and a further guide is given by the marks allocated. Thus, two marks indicate two main points to be given, and so on.

Some examination boards set structured short-answer questions without the spaces provided and it is up to the candidate to judge and present the answers very briefly to the correct length, e.g. structured question 4, Chapter 2.

(c) Structured Free-response Questions

This type of question demands a short piece of extended prose as an answer to a part of the question. Diagrams should be given if they help to clarify the answer and they can be drawn quickly. The question is usually in several parts, and the number of marks allocated gives a guide to length and the number of main points to be considered.

1.6 Writing extended prose

GCSE papers will not demand long essay-type answers. As stated in Section 1.5, the structured free-response question will be broken up into short answers. These can be compared to a short essay, although they may only carry 8 marks, for example, instead of 20 or 25. Nevertheless, care must be taken in planning and developing this type of answer and attention must be paid to the following points.

1. Examine the question carefully and think out **a plan** of the major points that should be included. If you have time, list them on a piece of rough paper. This will ensure that the material is applicable to the question and that you have not left out important points.
2. Once completed, **check the plan** against the question to ensure accuracy in terms of the demand made (see Section 1.5).
3. Try and **write simple sentences** with accurate use of technical terms and write them in a logical manner.
4. Be as **concise** as possible. This especially applies to questions where a comparative account is required. Points of similarity and contrast should be described together as they arise, e.g. two organisms with comparable points A, B, C, D, E and a, b, c, d, e should be compared A with a, B with b, C with c, and so on. It is a poor answer that gives a complete description of one set of points followed immediately by a description of the other set, and where there is no attempt at any form of real comparison.
5. **Diagrams**, labelled or annotated, should be used **where relevant** and where they are demanded in the wording of the question. They should never duplicate what has already been written, but should add something to the account. Repetition is a waste of time, as marks will only be given once for a correct point. You must always consider the time element, for drawing diagrams can take a long time.
6. Where the question is divided up into several parts, answer the various parts in the correct order. Beware of the question that offers a choice within the question, e.g. 'Comment on three of the following'. Make sure that the correct number of parts is answered.
7. Support statements with **definite examples** where applicable, as this will add emphasis to the answer. Always give a named example where required.
8. Keep a sharp eye on the **time**. It is important to answer all the questions that an examination demands. This is a matter of self-discipline.

1.7 Relevance of drawings and diagrams

Drawings or diagrams are sometimes demanded in a structured short-answer or structured free-response question. Since these are drawn under examination conditions, they cannot be expected to emerge as 'works of art'. They must be done quickly within limited time, and therefore some revision and practice are essential.

Viewing an illustration in a textbook and attempting to reproduce it in an examination will inevitably result in a vague interpretation of what needs to be shown. Therefore, any practice which is undertaken must always attempt to simplify.

The drawing or diagram should, as far as possible, comply with the following criteria:

1. It should be of **reasonable size**.
2. It should show the **correct proportions**.
3. The parts should be **accurately positioned**.

4. The drawing **should not be too detailed** and, therefore, should not take too long to draw.
5. Ensure that the drawing does indeed help the text of the answer or conform exactly to the requirement of the question.
6. Label lines should **point exactly** to the item labelled.
7. Label lines must **not cross over** each other.
8. Labels may be expanded into short notes (annotated), but these must not be repeated again in the written text.
9. Include **a title**.
10. Draw in pencil. If any parts are differentiated by shading, it may be necessary to provide a key.

1.8 Suggested practical investigations or experiments

The investigations or experiments are listed in a similar order to that of the text material in Chapters 2 to 16.

Observations of the characteristics of Man and other animals.

Cells: microscopic examination of cheek cells. Prepared sections or 35 mm slides can be used.

Electron micrographs to show ultrastructure of cells.

Enzyme experiments: (1) amylase on starch; (2) pepsin on egg white; (3) effect of temperature and pH on enzymes.

Diffusion, including dialysis and osmosis, using dried fruit and potato and including the effects of different concentrations of salt solutions on red blood cells.

Food tests: starch — iodine solution; reducing sugar — Benedict's solution; protein — biuret test; fat — ethanol test; detection of vitamin C. Testing of common foods for presence of different classes of nutrients. Determination of energy value of a peanut by burning under a test-tube containing a known mass of water.

Examination of teeth of students and extracted teeth.

Use of cellulose tubing with enzyme and substrate to show movement of breakdown product through the tubing to compare with gut functions.

Structure of the heart of a sheep. Microscopic slides or 35 mm transparencies to show structure of arteries, veins and capillaries.

Making a blood smear (within local authority regulations).

Investigations of the effects of exercise on pulse rates.

Examination of the lungs, bronchi and trachea of a sheep. Suitable models of these structures using a bell jar and balloons.

Carbon dioxide in exhaled air and inhaled air by breathing out through lime water or a hydrogen carbonate indicator.

Determination of vital capacity and tidal volume.

Smoking machine with a suitable trap to demonstrate presence of tars.

Examination of the kidneys of a sheep.

Examination of a complete human skeleton or separate bones. Microscope slides or 35 mm transparencies of the structure of cartilage and bone. Burning bones to show organic content and dissolving in acid for inorganic content.

Examination of the structure of the eye of a bullock or sheep. Experiments on skin sensitivity and field of vision; determination of the blind spot by use of + and o; perception of sound and direction; range and sensitivity of hearing. Knee-jerk reflex and action of iris, using a torch.

Observation of photographs showing human development.
Experimental genetics, using available material.
Pasteur's experiment to show that bacteria are present in air.
Experiments and observations on different methods of preserving food.
Make a model sand filter and consider its use in purifying water. Visit a water works and sewage works.
Observation of Man's effect on the land.

2 Man's Relationship with Other Living Organisms

2.1 What is life? Characteristics of living things

1. Nutrition (see Chapter 4): All living organisms need food. Green plants make their own food by photosynthesis incorporating the energy of sunlight, i.e. **autotrophic nutrition**. Animals obtain food by eating other organisms, digesting them enzymatically and absorbing the breakdown products, i.e. **heterotrophic nutrition**.
2. Respiration (see Chapter 7). All living organisms obtain energy from food when it is broken down. This process requires oxygen (aerobic respiration). Some energy can be released by a breakdown process without oxygen (anaerobic respiration).
3. Excretion (see Chapter 8). All living organisms excrete waste materials produced as a result of living processes (metabolism).
4. Response (see Chapters 10 and 11). All living organisms show irritability, i.e. the capability of responding to changes in the internal and external environment.
5. Reproduction (see Chapter 12). Every living organism has a limited life span. The organisms ensure species survival by producing new individuals with the same general characteristics as themselves.
6. Growth (see Chapter 12). All living organisms form new tissue from their food intake.
7. Movement (see Chapter 9). Animals are distinguished from plants by their ability to move from place to place (locomotion). This is essential in seeking out food and a mate.

All of these seven characteristics are observable qualities of the one, main, fact that **living organisms can extract, convert and use energy from their environment**, and thus maintain and even increase their energy content. In contrast, **dead organic material and non-living organic material tend to disintegrate, so that their energy content decreases.**

2.2 Man as a living organism

Man displays all of the above characteristics and is thus part of the animal kingdom. All animals are classified so that they can be placed with their closest relatives in clearly defined groups. The classification of Man is shown in Table 2.1.

Table 2.1   The classification of *Homo sapiens* (Man)

Taxonomic group	Name	Notes
Kingdom	Animal	Generally compact, mobile organisms, without the ability to make their own food
Phylum	Chordata	Includes both vertebrates and certain more primitive animals, such as sea squirts
Sub-phylum	Vertebrata	Chordates with backbone and skull
Class	Mammalia	Warm-blooded, hairy animals that give birth to live young, which they suckle
Order	Primates	Includes Old World and New World monkeys, and apes. Great development of brain and relatively large skull. Sight good: smell poor. Long growth period (12 years in apes, 17 in Man), nails and opposable digits. Development of forelimb for exploring environment. Usually only one offspring per birth

Man belongs to the mammals, which:

1. Are warm-blooded (homeothermic), i.e. having a constant body temperature.
2. Are covered with hair.
3. Possess mammary glands to suckle their young.
4. Are born alive after a gestation period.

Within the **class mammals**, Man belongs to **the Primates**. All varieties of Man (black, white, yellow, brown) belong to the one species, ***Homo sapiens***, which has the following features:

1. Well-developed forebrain and therefore a marked intelligence.
2. Thumb opposable to fingers and therefore used in grasping tools.
3. Rotating bones in the forearm used in turning tools.
4. Stereoscopic, colour vision.
5. An upright stance on the hind legs.
6. An extended, learning, period of childhood.
7. During this learning period develops a language.
8. Has developed a social organisation and considerable interdependence within that society.

Man's great **advance has been in cultural development**, so that humans are the most 'successful' animals. The reasons for this are:

1. Man is a **relatively unspecialised** animal in physical terms. Thus, humans cope well with climatic changes from the Arctic to the Equator. Furthermore, they have no particular food specialisations.
2. The **relatively great size of the forebrain** (15 cm) and its surface complexity gives a high level of intelligence. This adaptation together with individual and group memory have contributed to the development of child rearing, means of travel, building homes, developing agriculture, industrial advance and technology.
3. **Language** which can communicate simple and abstract ideas. Added to this, writing which allowed experience and knowledge to be passed from generation to generation.
4. The **industrial revolution in the nineteenth century** has led ultimately to the harnessing of power sources, advances in communication and other technology that we have today.

Table 2.2 Differences between green plants and animals

Green plants	*Animals*
1. Contain chlorophyll, enabling photosynthesis	Do not contain chlorophyll – no photosynthesis
2. Autotrophic – synthesise own food	Heterotrophic – must be provided with synthesised food
3. Show very limited movement but no locomotion	Show locomotion for seeking: (i) food (ii) shelter (iii) mate
4. Body often branched; no fixed shape, with unlimited growth	Body compact; fixed shape, with limited growth
Plant cells	*Animal cells*
5. Chloroplasts present containing chorophyll	No chloroplasts – no chlorophyll
6. Cell wall – made of cellulose – often strengthened by other substances, e.g. lignin	No cell wall
7. Single large cell vacuole	Small scattered vacuoles

2.3 Animals and plants

Plants synthesise their own food incorporating light energy (**photosynthesis**). This process is summarised as follows:

$$6CO_2 + 6H_2O \xrightarrow[\substack{\textbf{light} \\ \textbf{chlorophyll}}]{\text{in presence of}} C_6H_{12}O_6 + 6O_2$$

carbon dioxide + water → glucose + oxygen

The glucose is used (a) to be broken down by the plant to provide energy; (b) to be built up into more complex carbohydrates; (c) with the addition of nitrogen and other elements, to be built up into proteins.

The following process of aerobic **respiration** occurs in plants and animals:

$$C_6H_{12}O_6 + 6O_2 \longrightarrow 6CO_2 + 6H_2O + \text{energy}$$

The energy released in plants is used in chemical reactions, whereas in animals it is also used in movement, maintaining temperature levels, mating and numerous other activities.

Thus, it can be seen that animals and plants are completely interdependent.

1. Plants, by photosynthesis, **provide oxygen for animal respiration** and animals by this respiration **provide carbon dioxide for photosynthesis in green plants**.
2. Plants **provide carbohydrates, fats and proteins** for feeding herbivorous animals.
3. Animals by their death and decay **produce inorganic salts** which are **taken up by plants** and used in synthesis of organic compounds.

Thus, all life is cyclic.

Figures 2.1–2.4 show four natural cycles for nitrogen, carbon, water and organic/inorganic compounds. Note that they are drawn in different ways, using words only, incorporating symbols, clearly recognisable cycles and with an environmental background. All of these variations may be encountered in examination questions.

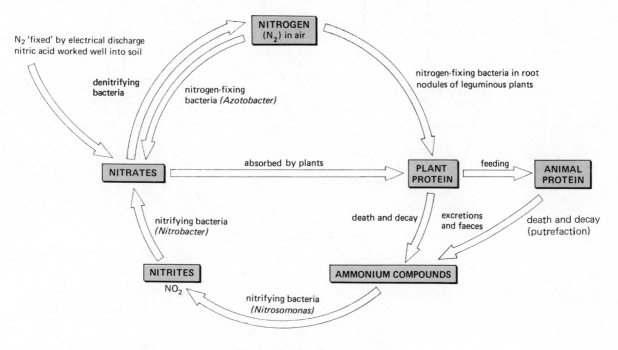

Fig. 2.1   The nitrogen cycle

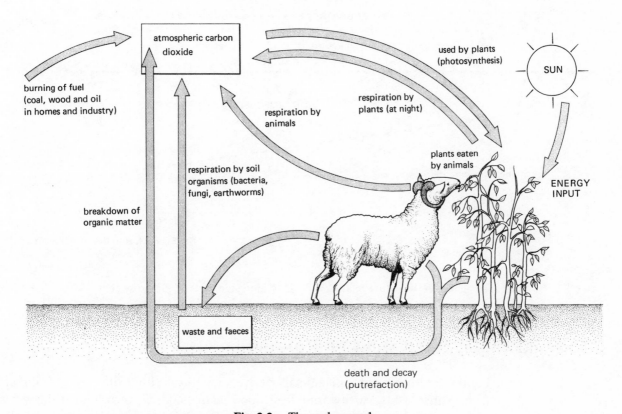

Fig. 2.2   The carbon cycle

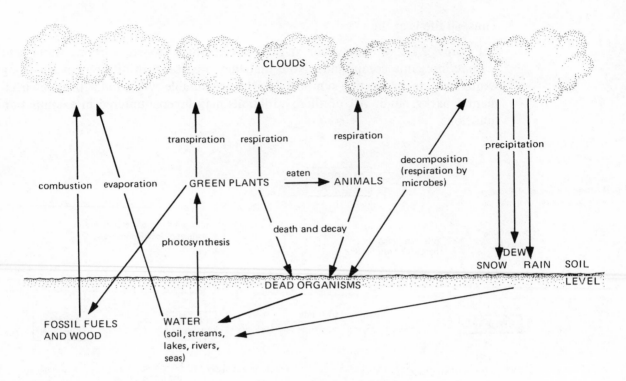

Fig. 2.3 The water cycle

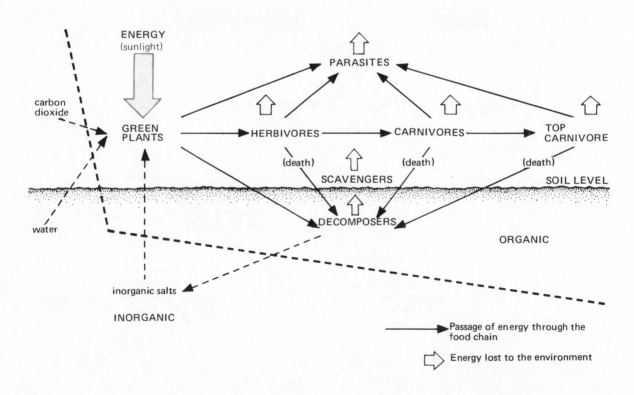

Fig. 2.4 The cycle of nature relating organic to inorganic parts of the ecosystem

2.4 Food chains, food webs and pyramids

Figure 2.5 shows that all animals derive their food either directly or indirectly from plants. Thus, earthworms feed upon plant material, thrushes eat the worms and sparrowhawks prey upon thrushes (see Fig. 2.6). Such a relationship is called

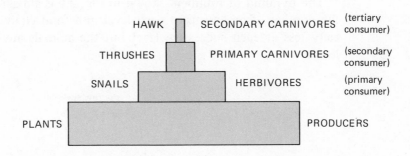

Fig. 2.5   A simple food chain and pyramid of numbers

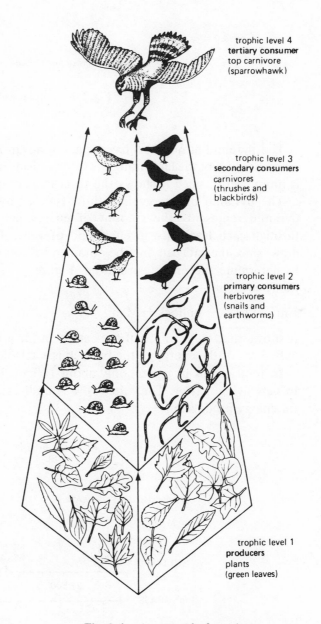

Fig. 2.6   A pyramid of numbers

13

a **food chain**. An interlinking series of food chains is called a **food web**. A food web based on Fig. 2.6 is shown in Fig. 2.7.

The **pyramid of numbers** shown in Fig. 2.6 is a diagrammatic representation of the number of organisms at each level in a food chain. The numbers are **numerically less** at each successive level but the **animals are larger** than those at lower levels.

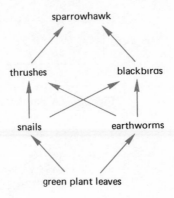

Fig. 2.7   A food web

The pyramid of numbers has certain disadvantages if one attempts to draw it to scale, because of the large numbers of organisms involved. The **pyramid of biomass** is more useful, since it shows the total mass of organisms (biomass) at each level.

The **pyramid of energy** is a third type of pyramid, in which each bar of the pyramid represents the amount of energy per unit area or volume that flows through each level in a given period of time. This pyramid is more difficult to draw, since the data can only be obtained by combustion of representative samples to determine the energy produced. Figure 2.8 shows the pyramid of energy for the food chain:

$1 \ m^2$ of grass → bullock → Man

It shows that only 16.3% of the energy in the grass is available to the bullock and only 3.5% of the energy in the bullock is available to Man.

In food chains in general only about 10% of the food matter eaten is converted to new living matter. The other 90% is undigested (passing out as faeces) or used for energy production.

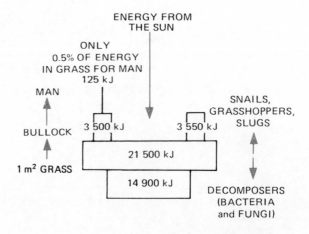

Fig. 2.8   Energy flow through a food chain involving Man and his food

2.5 World food production

From the figures for energy transfer given above, the conclusions must be:

1. The shorter the food chain the more efficient the energy conversion.
2. Eating plant material would support many more people than feeding plants to sheep and cattle and then eating the meat.

For these reasons the poorer people of the world must eat plant material (cereal grains, storage tubers, etc.), since they cannot afford to waste the energy available from their agricultural land by feeding their plants to animals.

There is a world food problem, because more than half of the people are under-nourished. Periodic famines throughout history have occurred in every part of the world. At present the belt of countries south of the Sahara in Africa are suffering extremes of starvation. The real problem is that population outruns food supply.

What methods can be used to provide food or reduce population?

1. Population control by using family planning techniques.
2. Intensification of agriculture giving increased yields:

   (a) genetic manipulation to give new high yielding varieties of crop plants;
   (b) irrigation;
   (c) provision of artificial fertilisers;
   (d) provision of plant crops with high protein yield to replace animal protein;
   (e) reclamation of land;
   (f) sea farming;
   (g) pest control of crops and storage areas.

3. New methods of farming:

   (a) cropping of game animals to produce protein;
   (b) culture of algae in shallow ponds;
   (c) cultivation of yeasts and other fungi;
   (d) hydroponic culture of food plants.

4. Education — the real key to better food production is to ensure that peasant farmers throughout the world are informed of new methods of agriculture.

2.6 Questions and answers

(a) Multiple-choice Questions

1 Which one of the following is characteristic of all living things?
   A breathing
   B excretion
   C feeding
   D locomotion
   E photosynthesis

2 Which one of the following is a mammalian characteristic in Man?
   A claws
   B hair
   C scales
   D feathers
   E moist skin

3 Which one of the following returns the nitrogen in animal cells to the air?
   A respiration
   B photosynthesis
   C transpiration
   D bacterial action
   E combustion

4 To which of the following animal groups does Man belong?
   A arthropods
   B marsupials
   C reptiles
   D mammals
   E amphibians

5 Which one of the following processes removes carbon dioxide from the air?
   A combustion
   B respiration
   C photosynthesis
   D fermentation
   E transpiration

6 Which one of the following terms describes the change from plant proteins to ammonium compounds?
   A nitrification
   B putrefaction
   C combustion
   D denitrification
   E fermentation

7 All organisms need nitrogen, but which of the following can produce nitrogen compounds from nitrogen gas?
   A bacteria
   B green plants
   C protozoa
   D fungi
   E algae

8 One feature that helps to show that Man is a mammal is that
   A the body has a blood system
   B the body has muscles enabling movement to take place
   C the young are fed on milk
   D the body has an endoskeleton
   E breathing is by means of lungs   (SREB)

9 One of the seven characteristics of living organisms is that they
   A move by walking
   B think
   C have a constant temperature
   D respire
   E have hair   (SREB)

10 Which of the following occurs when a green plant carries out photosynthesis?
   A water is released
   B carbon dioxide is used up
   C the energy comes from the breakdown of sugar
   D oxygen is required
   E chlorophyll is broken down   (NISEC)

(b) Structured Questions

1 Insert the most appropriate words, taken from the list below, into the spaces of the following passage:
   air; carbohydrates; carbon dioxide; carbon monoxide; cells; chlorophyll; energy; fats; heat; nitrogen; oxygen; water

16

In photosynthesis, the gas . . . . is absorbed from the . . . . through the stomata of the leaves and into the leaf . . . . This gas is used with . . . . and . . . . from the sun to make simple . . . . The process can only take place in the presence of the pigment . . . . Another gas, ..., is given off as a waste product. **(8)**

(OLE)

2 (a) Give three characteristics of animals that distinguish them from plants. (three lines)
(3)

   (b) Give three characteristics of mammals that distinguish them from other animals. (three lines)
(3)

(OLE)

3 (c) The diagram below shows the water cycle.

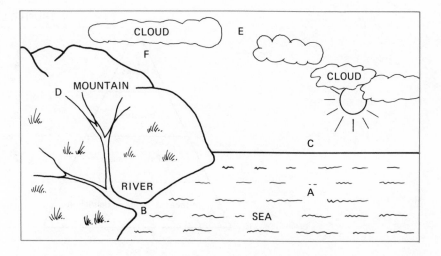

   (i) Using the letters **A** to **F** on the diagram of the water cycle, give the correct sequence of the water which is circulating, starting with **A**. **(2)**

   . . . .

   (ii) What is happening at the points marked **C** and **F**?

   C . . . .

   F . . . .

(2)
[Part question] **(YHREB, 1985)**

4 (a) The diagram represents the carbon cycle.

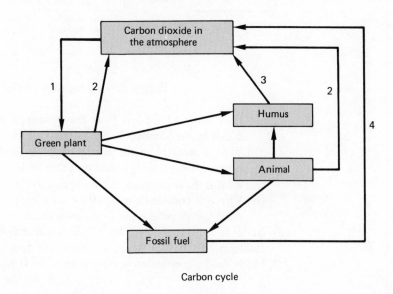

Carbon cycle

(i) Name the processes labelled **1, 2, 3** and **4**. **(4)**

(ii) Explain concisely how carbon dioxide entering a palisade cell in the leaf of a green plant becomes part of a glucose molecule. **(5)**

(iii) Name **two** groups of organisms that play a major role in process **3**. Where does this process normally occur? **(3)**

(iv) Name **one** common fossil fuel. **(1)**

(b) Explain how starch in a piece of potato becomes glucose and reaches the liver.
[See Chapter 5] **(10)**

(c) Explain how the provision of energy for the contraction of a muscle cell involves each of the following: ADP, enzymes, glucose, mitochondria and oxygen. **(7)**
[See Chapter 7] **(AEB, 1985)**

5 The diagram shows a simplified pyramid of biomass.

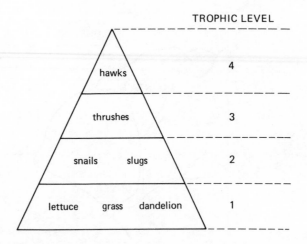

(a) State which trophic level contains the producers. Level. . . . .

(b) State which trophic level contains the secondary consumers. Level . . . .

(c) At which trophic level does holophytic nutrition occur? Level . . . .

(3)

(AEB, 1984)

6 (c)

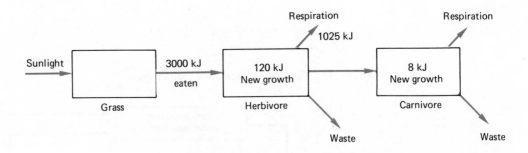

Energy flow through a food chain

(i) How many kJ are lost from the system (chain) through waste products produced by the herbivore?

(ii) What percentage of the energy available to the herbivore is converted into a form capable of being passed on to the carnivore?

(iii) What is the importance of chlorophyll to the above system?

(iv) Why is it considered more efficient for man to obtain his energy supply from a plant source rather than an animal source? **(6)**

(d) About two-thirds of the world's population are existing on a diet which is considered inadequate for healthy living. The Food and Agricultural Organisation was set up to tackle food problems on a worldwide scale. It recognised that in the long-term it was much better to spend money on educating people in agricultural methods and principles of nutrition than it was to supply shiploads of grain.

  (i) Give THREE different methods by which world food production could be increased.

  (ii) Suggest TWO reasons why education is a better long-term policy than importing foods whenever necessary. **(5)**

[Part question] **(NISEC)**

7

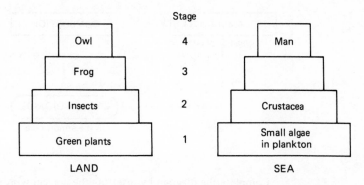

LAND   SEA

The diagram above represents examples of food chains in the sea and on the land.

(a) What is the ultimate source of energy in both food chains represented above? (one line) **(1)**

(b) (i) Name the process by which stage 1 makes carbohydrate. (one line) **(1)**

  (ii) What gas is absorbed by land plants for this process? (one line) **(1)**

(c) Name an organism which could occupy stage 3 in the food chain in the sea. (one line) **(1)**

(d) What use is made of the tissues of an insect after being eaten by a frog? Give two examples. (two lines) **(2)**

(L)

8 The diagram represents an outline of the nitrogen cycle.

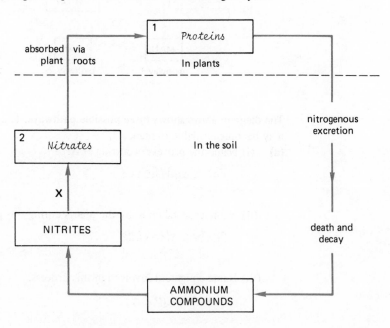

(a) Complete the diagram by inserting the correct words in the two numbered boxes. **(2)**

(b) Which group of organisms is responsible for the change labelled X on the diagram?

    Nitrifying bacteria

....

(1)

(AEB, 1983)

9 The diagram shows in outline the oxygen cycle in nature.

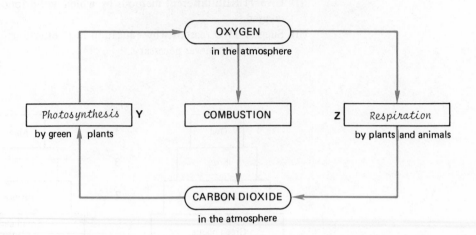

Complete the diagram by inserting the correct words in the spaces **Y** and **Z**. **(2)**

(AEB, 1983)

10

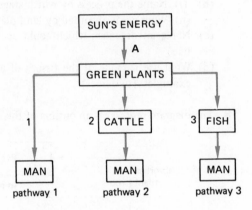

The diagram above shows three possible pathways, 1, 2 and 3, by which the sun's energy may become available to man.

(a)   (i) Name the process occurring at **A**.

Photosynthesis
....

(1)

(ii) Name a gas taken in by the plants during this process.

Carbon dioxide
....

(1)

(iii) Name a pigment involved in this process.

Chlorophyll
....

(1)

(b) Complete the table below by placing a tick in the most appropriate box. (There should be only one tick in each horizontal row.)

	Pathway 1	Pathway 2	Pathway 3	It is equal for all three
(i) Most energy would be available to man by means of	✓			

	Pathway 1	Pathway 2	Pathway 3	Pathways 2 and 3
(ii) Starving populations are most likely to have diets involving	✓			

	Pathway 1	Pathway 2	Pathway 3	There is equal loss in all three
(iii) Most energy in the form of heat is lost in		✓		

	Pathway 1	Pathway 2	Pathway 3	Pathways 1 and 3
(iv) If food shortage became a major problem in Britain, the best way of providing an adequate diet would be by means of				✓

(4)

(L)

11 The diagram below shows a simple food chain.

PRODUCER   example. . . . . *Algae OR Meadow grass*

↓

PRIMARY CONSUMER   example. . . . . *Roach*   *Bullock*

↓

SECONDARY CONSUMER   example. . . . . *Pike*   *Man*

(a) Write an example for each stage on the lines provided. (3)

(b) (i) What is the source of energy for producers?

. . . . *Sunlight* . . . .

(1)

(ii) How do consumers differ from producers in their source of energy?

Consumers obtain energy by eating producers or consumers; producers incorporate the energy of light.

(2)

(c) Why might it help the world food problem if humans ate less meat and more plant material?

By eating plants less energy is wasted in the food chain; about 5% of plant energy gained by herbivores, about 0.25% by carnivores.

(2)

(L)

(a) The diagram shows a food web.

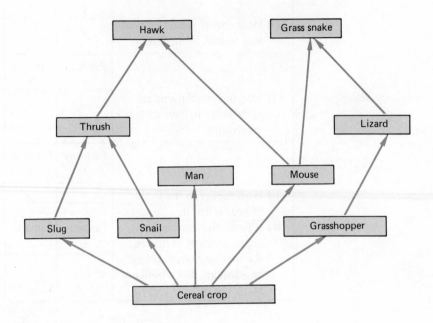

(i) Select from the food web a complete food chain consisting of **four** organisms. **(1)**

(ii) Explain how the energy from the sun is made available for man through this food web. **(5)**

(iii) Construct a diagram to show the circulation of carbon in nature. Using examples from the food web, incorporate *named* examples of a producer, a primary consumer (herbivore) and a secondary consumer (carnivore). **(8)**

(iv) Name a class of organisms, *not included in the food web*, which are necessary for the complete circulation of carbon in nature. **(1)**

(v) Explain what is likely to happen to the concentration of oxygen in the air surrounding the plants of a cereal crop during a 24-hour period of the growing season. **(7)**

(b) The element nitrogen is absorbed from the soil by plants.

(i) In what form is this element absorbed? **(1)**

(ii) Name a compound containing nitrogen which is made within the plant. **(1)**

(iii) Explain why the amount of nitrogen in the soil normally remains constant in nature. **(6)**

(AEB, 1984)

Answer

(a) (i) Cereal → snail → thrush → hawk

(ii) The cereals absorb the energy of sunlight during the process of photosynthesis. Carbon dioxide from the air and water from the soil are formed into simple carbohydrate (glucose) in the presence of chlorophyll. The energy of sunlight is incorporated into the glucose and this is built up into more complex compounds such as sucrose (cane sugar) and starch. These compounds are stored in the ripe fruit (grain) and Man consumes these organic substances when the grain is made into flour. When the sugars and starch have been assimilated into the body of Man, the process of respiration releases the energy in the food.

(iii)

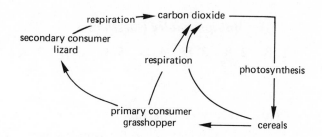

(iv) Decomposers.
(v) During daylight hours, when the process of photosynthesis is proceeding, carbon dioxide is being used up and oxygen is released. The concentration of oxygen therefore increases between and around the plants. On a warm day, with little breeze, this could rise above the average 21% for oxygen in the atmosphere.

Respiration in green plants is very slow and therefore, although some oxygen is used and carbon dioxide is released by plant cells in the light, the net exchange by day is to decrease the carbon dioxide and increase the oxygen. In the hours of darkness, however, respiration occurs without photosynthesis, with the result that some of the oxygen from the surrounding air is used up and carbon dioxide is released.

(b) (i) Nitrate.
(ii) Amino acids/protein.
(iii) The death of plants and animals in nature results in the decay of these organisms due to bacteria and fungi (decomposers). The proteins of the dead organisms are changed into ammonium compounds. Nitrifying bacteria then act on these compounds and change them into nitrites and nitrates. These substances are absorbed into plants through the root system and are used to synthesise amino acids and proteins. The full cycle of events has now restored the nitrogen of the plant proteins, and the cycle will turn again as the plants die. The plants could also be eaten by animals, so that the nitrogen passes through the animal proteins before being returned to the soil as urine or dead animal protoplasm.

Nitrogen of the air can be incorporated into plant material by nitrogen-fixing bacteria in the soil and in legumes. These plants, used as food by animals, can again return the nitrogen to the soil by animal excretion or death. Some nitrate can, however, be lost from the soil by the action of denitrifying bacteria.

All of these processes work together to keep the nitrogen content of the soil at a constant level.

Notes

1. This question is an example of a structured question gaining 30 marks and including three sections for extended writing. These three sections score more than two-thirds of the marks.
2. Where only one mark is given for a name, as in sections (a) (iv) and (b) (i), no time should be wasted in writing out a complete sentence.

(i) **Multiple-choice Questions**

1. B   2. B   3. D   4. D   5. C   6. B   7. A   8. C   9. D   10. B

(ii) **Structured Questions**

1. Carbon dioxide; air; cells; water; energy; carbohydrates; chlorophyll; oxygen
2. (a) Locomotion; holozoic/heterotrophic nutrition; fixed shape; limited growth; any three
   (b) Hair; mammary glands; diaphragm
3. (c) (i) **ACEFDB**
   (ii) **C**, evaporation; **F**, rainfall
4. (a) (i) **1**, Photosynthesis; **2**, respiration; **3**, putrefaction; **4**, combustion
   (ii) By the process of photosynthesis. The two raw materials carbon dioxide and water, in the presence of light and chlorophyll in the leaf, are changed into glucose.
   (iii) Fungi and bacteria bring about the breakdown process and this occurs in the soil.
   (iv) Coal (or petrol)
   (b) Starch is first chewed in the buccal cavity. Some starch is digested by salivary amylase. The starch is changed to maltose. The acidity of the stomach stops this process but it is continued in the intestine by amylase from the pancreas. The maltose produced is digested by maltase to the final breakdown product of starch, i.e. glucose. Glucose enters the bloodstream through the capillaries of the villi of the small intestine. All the blood from the intestines is collected into the hepatic portal vein and taken to the liver.
   (c) Oxidation of glucose molecules releases energy. This energy is incorporated into ATP as shown by the following equation:

   ADP + phosphate + energy = ATP

   All of these processes take place in the mitochondria with the aid of enzymes.
5. (a) Level **1**
   (b) Level **3**
   (c) Level **1**
6. (c) (i) $3000 - (1025 + 120) = 1855$ kJ
   (ii) $120 \div 3000 \times 100 = 4\%$
   (iii) Chlorophyll absorbs the energy of sunlight during photosynthesis in the grass plants.
   (iv) 3000 kJ of energy would be available from a plant source, whereas only 120 kJ would be obtained from eating a herbivore (such as a sheep). This is considering Man as a carnivore rather than as a herbivore.
   (d) (i) 1. Increased supply of artificial fertilisers.
       2. Use methods of farming to stop erosion of soils.
       3. By breeding methods develop better-producing crops.
   (ii) 1. Teach improved methods of agriculture appropriate to each area of the world.
       2. Food supply by outside agencies would breed dependency and not develop self-help in food production.

7. (a) Sunlight
  (b) (i) Photosynthesis
      (ii) Carbon dioxide
  (c) Fish
  (d) Protein for growth and fats for energy

Questions 8–11 have the answers supplied with the questions.

3 Cells, Organ Systems and Solutions

3.1 Cells

The cell is the basic unit of living organisms. When viewed under the light microscope ($\times 500$), all living animal cells have a similar structure (Fig. 3.1). There is a living outer **plasma membrane** enclosing a mass of **cytoplasm** in which is present a **nucleus**. The membrane is the outer boundary of the cell which controls the movement of materials in and out of the cell. Cytoplasm appears to be a semifluid, structureless substance containing a number of different solid particles. A number of droplets are also often visible, which may join together to form a larger **vacuole**.

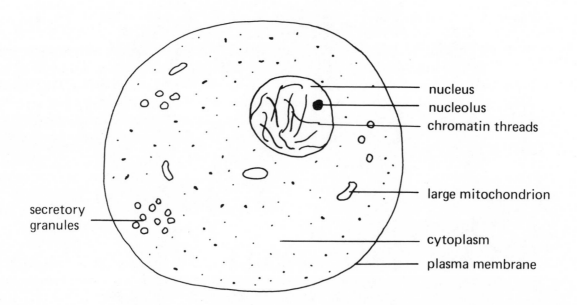

secretory granules

nucleus
nucleolus
chromatin threads

large mitochondrion

cytoplasm
plasma membrane

Fig. 3.1   Typical animal cell as seen with a light microscope

The nucleus is surrounded by a **nuclear membrane** and appears much darker than the surrounding cytoplasm in most preparations for the light microscope. When the cell is dividing, a number of dark-staining thread-like structures can be seen in the nucleus. These are **chromosomes**, which carry the genetic information of the organism.

There is aggregation of living material as follows:

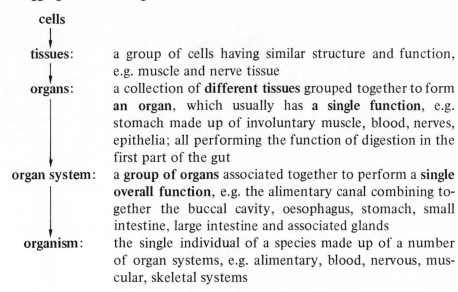

cells

tissues: a group of cells having similar structure and function, e.g. muscle and nerve tissue

organs: a collection of **different tissues** grouped together to form **an organ**, which usually has **a single function**, e.g. stomach made up of involuntary muscle, blood, nerves, epithelia; all performing the function of digestion in the first part of the gut

organ system: a **group of organs** associated together to perform a **single overall function**, e.g. the alimentary canal combining together the buccal cavity, oesophagus, stomach, small intestine, large intestine and associated glands

organism: the single individual of a species made up of a number of organ systems, e.g. alimentary, blood, nervous, muscular, skeletal systems

3.2 Tissues

There are four major types of tissues.

(a) Epithelial

Epithelial tissue is arranged in single or multi-layered sheets and covers the internal and external surfaces of the body of Man. **Simple epithelium** can be flattened, columnar, cubical or ciliated. It lines the cheek, alveoli and parts of kidney tubules (flattened) and the ducts of glands such as the pancreas (cubical); it forms part of the thyroid and gall bladder (columnar); and lines oviducts and respiratory passages (ciliated). **Compound epithelia** (stratified epithelia) forms the lining of the vagina and the epidermis of the skin.

(b) Connective

Connective tissue is the major supporting tissue of the body. It includes skeletal tissue. All connective tissue has fibres present in the form of yellow fibres (elastin) or white fibres (collagen). White fibres are tough and non-elastic, and are found in **white fibrous** connective tissue; they are present in tendons, cartilage, bone and the sclerotic layer of the eyeball. Yellow fibres are found in **yellow, elastic tissue** such as ligaments, lungs and associated air passages.

    Cartilage is a connective tissue with cartilage cells embedded in a matrix of chondrin. Cartilage may also have white or yellow fibres giving it properties associated with these fibres. **Bone** is a calcified connective tissue containing a matrix (30% organic material and 70% bone salts – principally calcium). Bone cells situated throughout the matrix are connected with blood vessels (see Fig. 3.3 and structured question 1).

    Blood is regarded as a connective tissue (see Chapter 6).

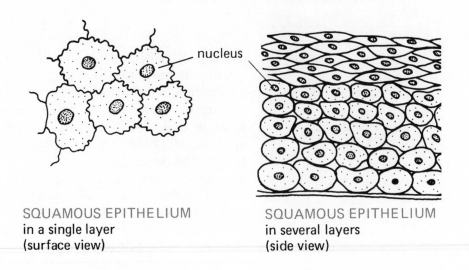

SQUAMOUS EPITHELIUM
in a single layer
(surface view)

SQUAMOUS EPITHELIUM
in several layers
(side view)

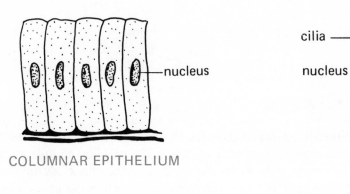

COLUMNAR EPITHELIUM

CILIATED
COLUMNAR
EPITHELIUM

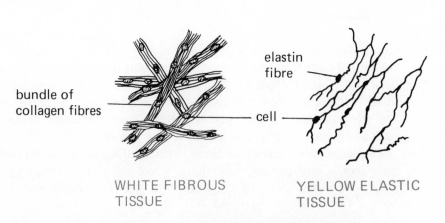

WHITE FIBROUS
TISSUE

YELLOW ELASTIC
TISSUE

Fig. 3.2 Types of tissues

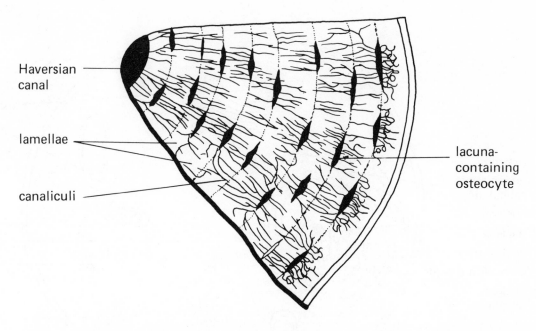

Fig. 3.3 Transverse section of bone — the lamellae are layers made up of magnesium and sodium, carbonates and nitrates

(c) Muscle

There are three types of muscle:

1. **Voluntary, skeletal or striated** — attached to bones and under the control of the will (cerebrum of the brain). These muscles provide powerful, rapid contraction.
2. **Involuntary** — found in the gut, blood vessels and bladder. These muscles provide slow, rhythmic contractions, as in the peristalsis of the gut.
3. **Cardiac** — found only in the heart, giving rapid, rhythmical contractions without showing any fatigue.

(d) Nervous

Nervous tissue consists of **neurons**, the functional units of the nervous system (see Chapter 11).

3.3 Ultrastructure

In recent years the structure of the cell has been more clearly demonstrated by the use of the electron microscope and special techniques. Much higher magnifications ($\times 500\,000$) can be obtained, revealing the following structures (see Fig. 3.4):

1. **Mitochondria** — these are the sites of **respiration**. Energy is released from carbohydrates and fats. The mitochondria are often rod-shaped, with an outer double membrane and projections of the membrane into a matrix.
2. **Ribosomes** — these are the sites of **protein synthesis**. They are very small organelles made of protein and RNA (ribonucleic acid). They can be free in the cytoplasm or bound to the endoplasmic reticulum.

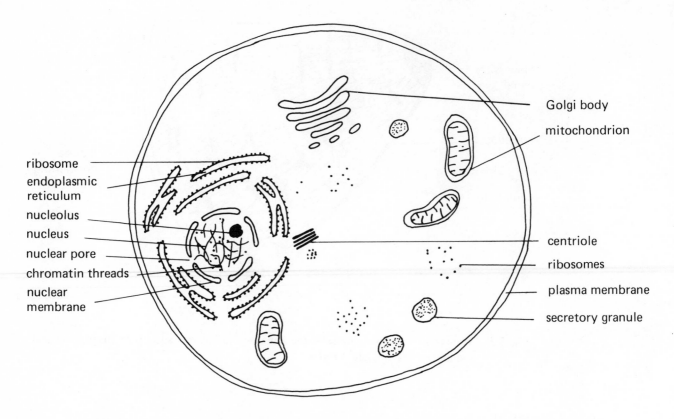

ribosome

endoplasmic reticulum

nucleolus

nucleus

nuclear pore

chromatin threads

nuclear membrane

Golgi body

mitochondrion

centriole

ribosomes

plasma membrane

secretory granule

Fig. 3.4   Ultrastructure of a generalised animal cell as seen with the electron microscope

3. **Endoplasmic reticulum** — this is the site of **lipid synthesis**. It consists of a series of flattened membrane-bounded sacs forming tubes and sheets. It is continuous with the outer plasma membrane of the cell.
4. **Nucleus** — this organelle has an envelope of two membranes perforated by nuclear pores. It contains chromatin, which forms chromosomes during cell division. It also contains a **nucleolus**. Chromosomes contain DNA (deoxyribose nucleic acid), the molecule of inheritance. DNA is arranged into **genes**, which control all activities of the cell. Nuclear division occurs, involving **chromosomes** (see Chapter 12). The nucleolus manufactures ribosomes.

3.4 Living processes in the cell

(a) Respiration

Energy is required for the movement of molecules into and out of the cell; for breaking down large molecules and building up new molecules; and for specific cell action, e.g. muscle contraction. The energy comes from sugars and fats, which are broken down, the energy present in their chemical bonds being transferred to other molecules in the cell. Aerobic and anaerobic respiration is summarised in Chapter 7. The site of all of these reactions is the mitochondrion.

(b) Transport

Transport within the cell and across membranes can be by diffusion and osmosis.
    Diffusion is the movement of molecules from a region of high concentration to a region of low concentration. It could account for the movement of glucose,

amino acids and oxygen into the cell as well as carbon dioxide and water out of the cell. This is sometimes called passive transport.

Active transport is the movement of molecules against the concentration gradient. It requires energy, and this is a measurable feature of active transport. A carrier substance is required to move the molecules across the plasma membrane.

Osmosis is a special case of diffusion: the diffusion of a solvent, water, from a high concentration of water (a weak solution) to a low concentration of water (a strong solution). The two solutions are separated by a membrane said to be differentially or selectively permeable. The solvent molecules (water) move in both directions across the membrane, but there is a net increase on the side of the strong solution, since more molecules move in that direction (see Fig. 3.5).

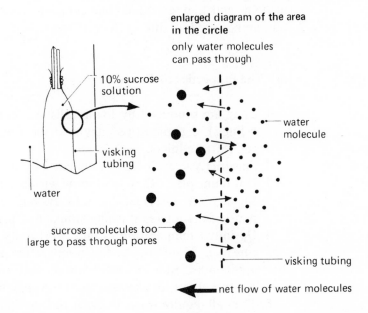

Fig. 3.5 Diagrammatic representation of the pores present in a membrane and the movement of the molecules

The cell plasma membrane is selectively permeable. The cell has a cytoplasm containing dissolved substances, and thus it will absorb water by osmosis from any weaker solution that surrounds it. If surrounded by a stronger solution, the cell will lose water. Therefore, the tissue fluids of the body must always be of the same strength as the cytoplasm. The tissue fluids are derived from the blood, and the latter is homeostatically controlled by the action of brain, kidneys and liver. The osmotic potential of the blood in the capillaries, together with its hydrostatic pressure, aids the exchange of water between the capillaries and the tissue fluids (see Fig. 6.2).

(c) Manufacture of Proteins

Amino acids are assembled into proteins at the ribosomes. The sequence of amino acids in each protein is determined by the molecules of ribonucleic acid (RNA). RNA is made in the nucleus and passes out through the nuclear pores into the cytoplasm.

31

(d) Metabolism

The living processes in the cell are collectively termed metabolism. The building-up processes are called **anabolism** (e.g. protein synthesis), while the breaking-down processes are called **catabolism** (e.g. respiration).

3.5 Enzymes

All anabolic and catabolic activities in the cell proceed with the help of enzymes. Without their action chemical reactions would occur very slowly. **Enzymes are proteins** and they act as **organic catalysts**, i.e. they speed up the rate of a reaction but are not used up during the process.

(a) The Properties of Enzymes

1. They are **produced by living cells**. Some function outside of the cell (extra-cellular), e.g. digestive enzymes (see Chapter 5); most function within the cell (intracellular).
2. **Specificity** – each type of enzyme speeds up one reaction only.
3. **Optimum pH** – they work only within a narrow range of acidity or alkalinity for any one enzyme, e.g. salivary amylase pH 7.5 in the buccal cavity and pepsin functions best in the stomach at pH 2.0.
4. **Optimum temperature** – they work only within a narrow range of tempera-ture, i.e. 35–40°C. As the temperature falls, they work more slowly, and above 40°C they become slower, until they are inactivated at about 60°C. At, and above, this temperature the enzyme is denatured by the heat.
5. They all **require water** in order to function.

3.6 Questions and answers

(a) Multiple-choice Questions

1 Which one of the following contains most adenosine triphosphate (ATP)?
   A ribosome
   B mitochondrion
   C nucleolus
   D chromosome
   E nucleus
2 In which one of the following is protein manufactured?
   A ribosome
   B mitochondrion
   C nucleolus
   D chromosome
   E nucleus
3 Which one of the following cells has no nucleus?
   A ovum
   B white blood cell
   C sperm
   D red blood cell
   E neuron

4 Which one of the following surrounds an animal cell?
   A cell wall
   B plasma membrane
   C skin
   D jelly
   E nuclear membrane
5 Which one of the following cells has the most mitochondria?
   A sperm cell
   B neuron
   C ovum
   D muscle cell
   E liver cell
6 Which one of the following is the only way in which plant cells differ from animal cells?
   A Plant cells all possess chlorophyll.
   B Plant cells have a cellulose cell wall.
   C Plant cells are larger.
   D Plant cells are not as specialised.
   E Plant cells are all living.
7 Which one of the following is present in the vacuole of a plant cell?
   A cytoplasm
   B endoplasm
   C protoplasm
   D ectoplasm
   E cell sap
8 Which one of the following is the part of the cell that controls a single inherited factor?
   A gene
   B chromosome
   C gamete
   D nucleus
   E ovum
9 The function of ribosomes in a cell is to
   A control substances entering the cell
   B hold chromosomes in place
   C release energy
   D act as genes
   E produce proteins   (SREB)
10 Which one of the following lists shows the correct order, starting from the smallest and finishing with the largest?
   A cells, organelles, organs, tissues, organ systems
   B organelles, cells, tissues, organs, organ systems
   C cells, organs, organ systems, tissues, organelles
   D organ systems, organs, organelles, tissues, cells
   E organelles, organs, cells, tissues, organ systems   (SREB)

(b) Structured Questions

1 The diagram below shows some cells from the human body.

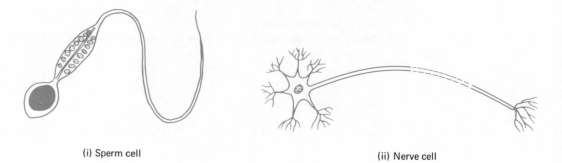

(i) Sperm cell   (ii) Nerve cell

(iii) Red blood cell

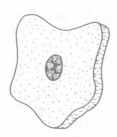

(iv) Epithelial cell

Using the diagrams to help you, briefly describe for each cell one way in which it is adapted for carrying out its particular function.

  (i) sperm cell   (two lines)
  (ii) nerve cell   (two lines)
  (iii) red blood cell   (two lines)
  (iv) epithelial cell   (two lines)

                                                   (4)
                                          (SREB)

2 (a) (i) The figure below is a diagram of a generalised animal cell.

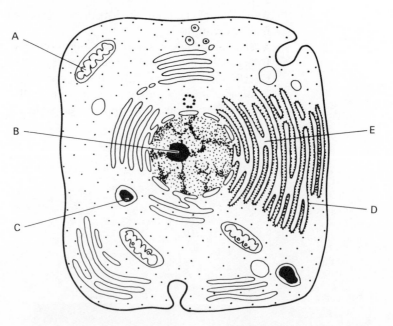

Name the structures labelled **A, B, C, D, E**, stating the principal function of each.

A, B, C, D and **E** two lines each   **(10)**

(ii) What is the significance of the foldings of the inner membranes of **A**?

(iii) Indicate with a tick (√) those cells in which you would expect to find relatively large numbers of **A**.

Tick

. . . . cartilage

. . . . muscle

. . . . conjunctival cells

. . . . cells lining the convoluted tubules of the nephrons

. . . . cells lining the pancreatic duct

. . . . interstitial cells of the testis   **(UCLES)**

3 The diagrams below show different types of tissues found in the body. For each of the diagrams **(a)** to **(e)**, (i) name the tissue and (ii) name an organ where it can be found.

(a)

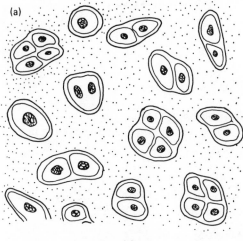

(i) . . . . *Cartilage* . . . .

(ii) . . . *Ends of long bones* . . . . . .

(2)

(b)

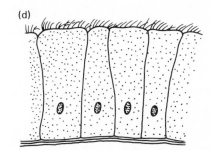

(i) . . . . *Heart muscle* . . . .

(ii) . . . . *Heart* . . . .

(2)

(c)

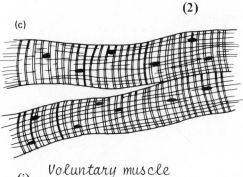

(i) . . . . *Voluntary muscle* . . . .

(ii) . . . . . *Leg muscles* . . . .

(2)

(d)

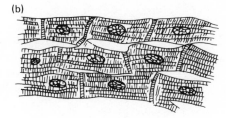

(i) . . . . *Ciliated epithelium* . . . . .

(ii) . . . . *Trachea* . . . .

(2)

(e)

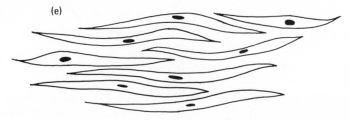

(i) . . . *Involuntary muscle* . . . .

(ii) . . . *Wall of alimentary canal* . . . . .

(2)

(L)

4 CELLS

Study the diagrams below showing three types of cells and answer the questions on them.

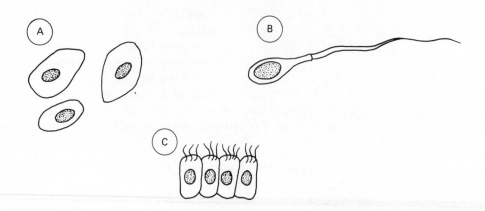

(a) Name cell B . . . . *Sperm* . . . . (1)

(b) Where are cells of type B produced? . . *Testis* . . . .(1)

(c) Give one place where cells of type C are to be found in the body.

. . . . *Oviduct* . . . . (1)

(d) What are the 'hairs' on cell C called? . . . *Cilia* . . . .(1)

(e) What is the function (job) of these 'hairs'?

. . . *To move fluids (mucus)* . . . . (1)

(f) Write down **TWO** structural similarities between all three cells.

1. . . *Nucleus* . . . .

2. . . *Plasma membrane* . . . .

(2)

(EMREB)

5 Below is a diagram of a generalised animal cell.

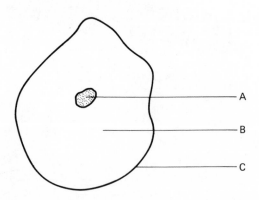

(a) Name parts A, B, C

A . . *Nucleus* . . . .

B . . *Cytoplasm* . . . .

C . . *Plasma membrane* . . . .

(3)

(b) Gases diffuse into and out of cells.

    (i) Name a gas normally diffusing into a cell . . . . *Oxygen* . . . .

    (ii) Name a gas normally diffusing out of a cell. . . . . *Carbon dioxide* . . . .

(c) Name the chemical process for which the gas diffusing into a cell is used.

    Respiration

    . . . .

                                    (3)

                                    (LREB)

6 Briefly describe a function of each of the following cell organelles.

(a) Mitochondrion

    It is the site of the process of respiration

    by which energy is released from carbohydrates.

                                    (2)

(b) Golgi body

    It is concerned with the transport of

    many cell materials such as enzymes from the E.R.

                                    (2)

(c) Ribosome

    It is the site of protein synthesis based on

    information from the nucleus in the form of RNA.

                                    (2)

(d) Nucleus

    It contains chromosomes. These contain DNA, which

    controls the activities of the cell.

                                    (2)

                                    (L)

7 The diagram represents some features of a cell as seen with the electron microscope.

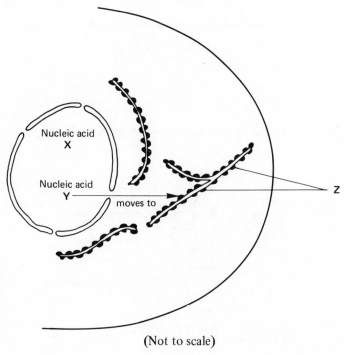

(Not to scale)

(a) Identify the nucleic acids present in the nucleus.
  (i) Nucleic acid **X**.

  DNA (deoxyribonucleic acid)
  ....
  (ii) Nucleic acid **Y**.

  RNA (ribonucleic acid)
  ....
(b) Name the structures **Z**.

  Ribosomes
  ....
(c) Briefly explain the relationship between the nucleic acids and the function of structure **Z**.

  DNA provides a code for the manufacture of protein. This
  ....
  code is conveyed by RNA to the ribosomes to make
  ....
  proteins. RNA moves through the nuclear pores.
  ....

(7)
(AEB, 1985)

8 The diagram represents a structure present in most cells.

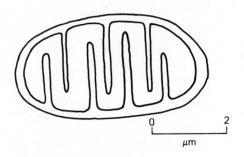

(a) Name this structure.   (one line)   **(1)**

  Mitochondrion
  ....
(b) State precisely its role in a cell.   (one line)   **(1)**

  Release energy by the process of respiration
  ....

(2)
(AEB, 1983)

9 The figure below represents animal cells and plant cells.

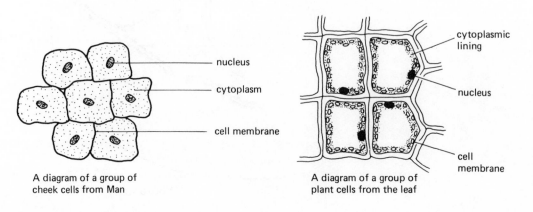

A diagram of a group of
cheek cells from Man

A diagram of a group of
plant cells from the leaf

(a) In the table below state **three** differences between the structures of the plant cells and the animal cells shown in the figure.

	Plant cells	Animal cells
1	Green chloroplasts present	No chloroplasts
2	Cell wall present	No cell wall
3	Large vacuole in centre of cell	No large vacuole

(b) Suppose that each group of cells was placed in a 5% sucrose solution. Describe briefly what would happen in each case.

(i) Animal cells

The cells would lose water by osmosis and shrivel up.

(ii) Plant cells

The cells would lose water by osmosis. The cytoplasm shrink away from cell walls; vacuoles become smaller. **(2)**

(c) Name, and briefly describe, the type of plant cell concerned with

(i) the transport of water and salts

Xylem vessels and tracheids. Dead cells with a continuous internal lumen for transport. **(1)**

(ii) the transport of sugars and amino acids

Phloem sieve tubes. Lining cells with each unit separated by a sieve plate. **(1)**

(UCLES)

Free-response Question

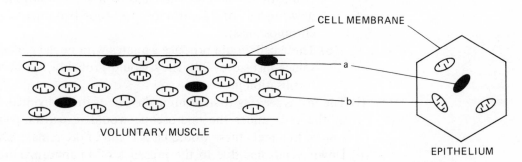

(a) (i) Name the structures a and b which are present in both the cell types shown.
    (ii) From your knowledge of its function how would you account for the great number of b type structures in the muscle?
    (iii) In what way is the structure of voluntary muscle different from most common human cell types? **(5)**

(b) The chromosomes present in two different types of cell in a healthy adult were counted, with the following result:

Cell Type A: 23 chromosomes

Cell Type B: 46 chromosomes

  (i) What explanation would you give for the result obtained for Cell Type A?

  (ii) Name ONE organ of the body in which Type A cells are formed.

  (iii) A few of the cells taken from the same organ as Cell Type A were found to have 92 chromosomes.

    (a) Explain why there was such a large number of chromosomes in these cells.

    (b) What relationship exists between these cells and Cell Type A? **(5)**

(c) Athletes taking part in international competition undergo a 'sex test' to confirm their eligibility to compete in men's or women's events. The test is performed by scraping a few cells from the inside of the cheek and examining a stained sample of them under a microscope.

  (i) What are the examiners looking for and how does the evidence provided by the test enable the correct decision on eligibility to be made?

  (ii) Identify ONE inherited disease which could be detected by this technique and indicate what result you would expect in this case. **(4)**

(d) It is estimated that in the United Kingdom one marriage in six is childless. Because of recent advances in medical knowledge and techniques, some of these childless couples are now able to have children by methods which replace or supplement the normal pattern of conception.

Outline the biological principles behind any TWO different techniques which have been successful in promoting such births. **(6)**

(NISEC)

Answer

(a) (i) a, nucleus   b, mitochondria

  (ii) Muscle contraction requires large amounts of energy. This is released by aerobic and anaerobic respiration through the activity of enzymes in the mitochondria and in the cytoplasm. Thus, one would expect large numbers of these structures in voluntary muscle.

  (iii) There is no definite individual cell structure with each cell having a nucleus. Each muscle fibre has numerous nuclei scattered along its length.

(b) (i) Cell type A must be a gamete, either sperm or egg, since it contains the haploid number of chromosomes for humans, i.e. 23.

  (ii) Testis or ovary.

  (iii) (a) As a result of nuclear and cell division, there has been replication of the chromosomes but no formation of clearly separate nuclei. Thus, a mitotic division of a nucleus with 46 chromosomes has produced a new nucleus of 92 chromosomes (not two nuclei each containing 46 chromosomes).

    (b) These cells could produce gametes with nuclei containing 46 chromosomes instead of 23. They would probably not be viable during any fertilisation process.

(c) (i) The test is seeking to determine the sex of the individual. The examination of the cells under the microscope would reveal the pair of sex chromosomes. In a male these would be XY and in the female XX.

  (ii) Down Syndrome due to the presence of an abnormal number of chromosomes. There is one extra chromosome, such that each nucleus has 47 chromosomes instead of 46.

(d) Some women are unable to ovulate, owing to hormonal imbalance. This monthly release of an ovum during the menstrual cycle is stimulated by the follicle stimulating hormone (FSH) released by the pituitary gland into the bloodstream. These women can be treated by the so-called 'fertility drug', which is in fact FSH obtained from the pituitary glands of other mammals.

Other women have a blockage of the oviducts, so that eggs released from the ovaries are unable to reach the lower end of the oviducts and thus be fertilised. If surgery fails to help in such cases, ova can be obtained from mature follicles in the ovaries and then fertilised by sperm in a test-tube. The father's sperm is used. The fertilised egg divides into several cells, and this small ball of eggs is transferred back into the uterus to continue its proper growth.

Infertility in males can sometimes be overcome by using the father's concentrated sperm. Otherwise donor sperm must be used, i.e. AID – artificial insemination by donor.

Notes

1. This is a structured free-response question. There is no indication of the length required for each answer except for the number of marks.
2. Part (b) (iii) is unusual in that it involves the concept of polyploidy, i.e describing a nucleus, cell or organism with three or more times the haploid number of chromosomes.

(c) Answers to Objective and Structured Questions

(i) *Multiple-choice Questions*

1. B   2. A   3. D   4. B   5. D   6. B   7. E   8. A   9. E   10. B

(ii) *Structured Questions*

1 (i) Long tail for swimming towards the egg.
   (ii) Long axon for the conduction of impulses.
   (iii) Haemoglobin and a large surface area for the pick up and transport of oxygen.
   (iv) Flattened shape to form, with other like cells, a protective outer layer of organs.
2 (a) (i) A, mitochondrion; respiration
            B, nucleolus; manufactures ribosomes
            C, vesicle; contains enzymes
            D, ribosome; protein synthesis
            E, endoplasmic reticulum; transports proteins
       (ii) To provide a large surface area for the site of multi-enzyme actions.
       (iii) A tick should be placed for the following cells: muscle; cells lining the convoluted tubules of the nephron; interstitial cells.

Questions 3–9 have the answers supplied with the questions.

4 Nutrition, Food and Diet

4.1 Classes of food

The process by which an organism obtains its food materials in order to carry out its vital functions is called **nutrition**. As far as Man is concerned, this is really a question of buying the right quantity and quality of food to form the **diet**. A **balanced diet** supplies optimum amounts of each type of nutrient. These amounts will vary with such factors as age, sex and occupation.

There are six classes of food:

1. Carbohydrates ⎫
2. Fats ⎬ required in large quantities for energy production.

3. Proteins — required in large quantities for growth.
4. Mineral salts — required in traces for vital processes (see Table 4.2).
5. Water — required as a solvent for chemical reactions.
6. Vitamins — required in small quantities to maintain health (see Table 4.1). Most vitamins cannot be manufactured by the body.

(a) Carbohydrates

Carbohydrates are the simplest carbon compounds, with carbon, hydrogen and oxygen in their molecules, the hydrogen and oxygen being in the same proportion as in water. Thus:

$$C + H_2O = C_n(H_2O)_m$$
$$\text{carbohydrate}$$

where n and m may be the same or different numbers.

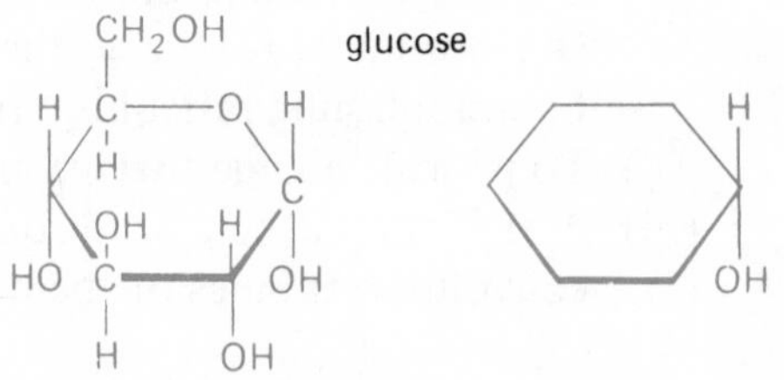

Fig. 4.1 Formula of glucose and simplified convention of a glucose molecule

Carbohydrates have a basic building molecule — the simple sugar or monosaccharide. This can be built up into more complex compounds:

monosaccharides — one molecule;
disaccharides — two molecules;
polysaccharides — many molecules.

Fig. 4.2 Simplified convention of a maltose molecule

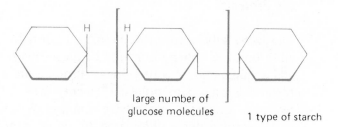

large number of
glucose molecules

1 type of starch

Fig. 4.3 Simplified convention of a starch molecule

Glucose is an example of a monosaccharide (formula $C_6H_{12}O_6$) and **ribose** is another example (formula $C_5H_{10}O_5$). **Sucrose** (cane sugar, beet sugar or table sugar: formula $C_{12}H_{22}O_{11}$) is formed by the joining together of two monosaccharides with the loss of a molecule of water. This method of joining together is called condensation. **Maltose** is formed from two glucose molecules, while **lactose** (milk sugar) is formed from a glucose molecule and another called galactose.

Polysaccharides are the largest carbohydrate molecules. They have two main functions: (1) as **energy source**, e.g. starch and glycogen; (2) as **structural molecules**, e.g. **cellulose** forms the cell walls of plants and, being indigestible, becomes roughage in the diet.

Plants manufacture glucose by the process of photosynthesis. This simple sugar is built up into storage materials such as sucrose, and starch. These substances are present in large quantities in:

cereal crops — wheat, oats, maize;
ground crops — Irish potatoes, sweet potatoes, yams, carrots, turnips;
leguminous crops — beans, peas, soya beans.

Man and animals use these food stores to provide their own carbohydrate requirements. If large quantities of carbohydrates are consumed by Man over and above his energy needs, the carbohydrates are changed into fats and stored in the body. In general, Man throughout the world has enough carbohydrate in his diet, except where drought or flooding results in famine.

(b) Lipids (Fats)

Lipids are organic compounds which contain carbon, hydrogen and oxygen, like the carbohydrates, but the proportion of hydrogen to oxygen is not the same as in water. The proportion of oxygen in relation to other elements is very low. The lipids include fats and oils, which are similar, except that **fats** are **solid** at **room temperature** and **oils** are **liquid** at room temperature. Plants tend to have liquid oils which contain unsaturated **oleic acid**, whereas animals store fats which contain **palmitic acid** ($C_{16}H_{32}O_2$) and **stearic acid** ($C_{18}H_{36}O_2$).

Fats are composed of two types of chemical — **fatty acid** and **glycerol**: such a combination is called a **glyceride**. Fats, like carbohydrates, are important sources of energy, and because of their high energy content and chemically inert nature

43

they are the most economical storage materials for living organisms. Humans eating more food than their bodies can use will accumulate fat. Animals living in cold climates have fat reserves under the skin to provide insulation against heat loss, e.g. polar bears and seals. Similarly, Eskimos have fat reserves and their small stature also aids heat retention.

(c) Proteins

The living substance, protoplasm, is essentially a solution of proteins in water. In addition to carbon, hydrogen and oxygen, all proteins contain **nitrogen. Phosphorus** and **sulphur** may also be present. Proteins are made of smaller units, amino acids, forming a long-chain molecule. Twenty-three different amino acids are found, and twelve of these are particularly common. When amino acids are joined together, a molecule of water is eliminated during the process. The bond between two amino acids is called a **peptide** bond. When many acids are linked together by this peptide linkage, the product is called a **polypeptide**. Polypeptides are joined to form the protein molecule, which is then known as a **polymer**. The protein **keratin**, which is present in hair, has sulphur bonds similar to those found in the insulin molecule (see Fig. 4.4). Combing the hair with a hot comb can break these bonds temporarily and thus straighten the hair. The bonds soon rejoin again and the hair becomes kinky.

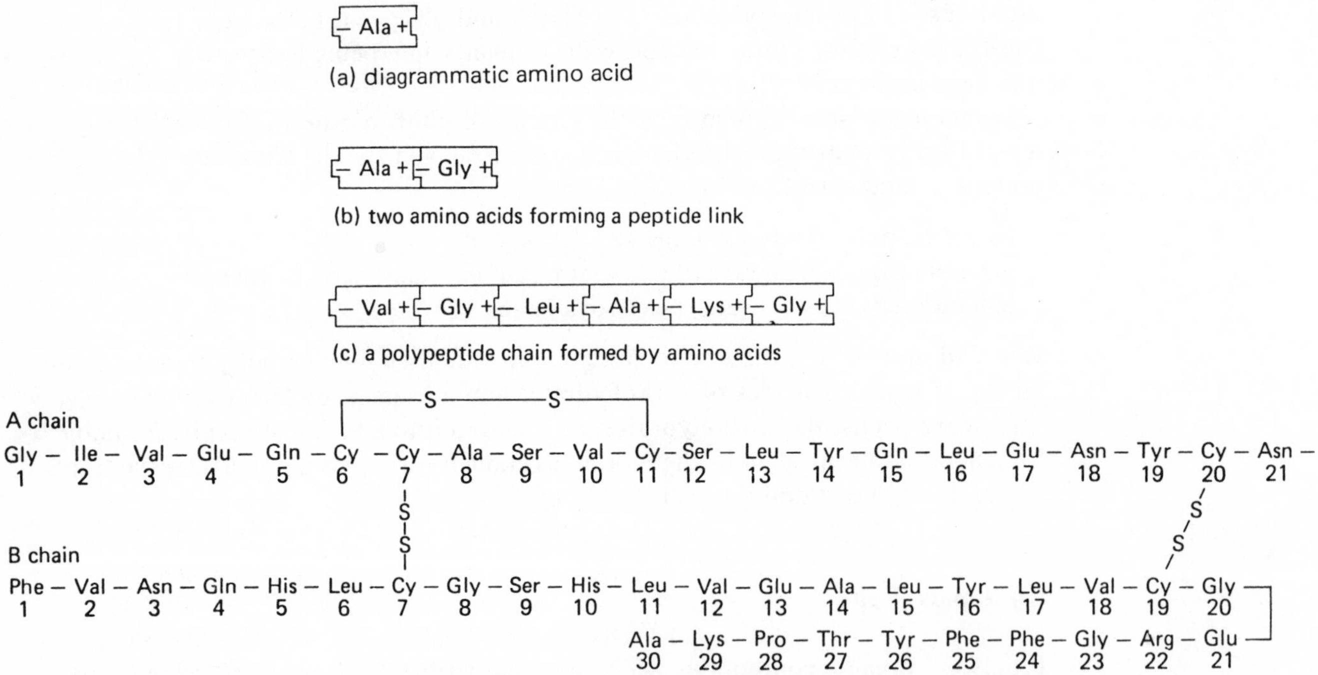

┌─ Ala +┌

(a) diagrammatic amino acid

┌─ Ala +┌─ Gly +┌

(b) two amino acids forming a peptide link

┌─ Val +┌─ Gly +┌─ Leu +┌─ Ala +┌─ Lys +┌─ Gly +┌

(c) a polypeptide chain formed by amino acids

A chain

Gly — Ile — Val — Glu — Gln — Cy — Cy — Ala — Ser — Val — Cy — Ser — Leu — Tyr — Gln — Leu — Glu — Asn — Tyr — Cy — Asn —
 1   2   3   4   5   6   7   8   9   10   11   12   13   14   15   16   17   18   19   20   21

B chain

Phe — Val — Asn — Gln — His — Leu — Cy — Gly — Ser — His — Leu — Val — Glu — Ala — Leu — Tyr — Leu — Val — Cy — Gly
 1   2   3   4   5   6   7   8   9   10   11   12   13   14   15   16   17   18   19   20

Ala — Lys — Pro — Thr — Tyr — Phe — Phe — Gly — Arg — Glu
30   29   28   27   26   25   24   23   22   21

(d) the sequence of amino acids in the insulin molecule

Fig. 4.4 Polymerisation, showing the amino acid sequence in an insulin molecule

Living processes are controlled by enzymes, each of which is a protein molecule. Enzymes and other proteins are very sensitive to heat, and when heated the protein molecules lose their special properties and become '**denatured**'.

4.2 Food

The reasons why the body needs food are:
1. for **growth** — the initial increase in cells in order to grow to adult size, and the repair of damaged or worn-out tissues;
2. for **energy** — to provide energy to drive the chemical processes, for mechanical work of muscles and for the maintenance of body temperature;
3. for **health** — to afford protection against disease and to provide raw materials for the manufacture of secretions such as hormones and enzymes.

4.3 Sources in diet of the main classes of food

1. **Carbohydrates** — potatoes, cereals, bread, sugar, rice, cassava, bananas.
2. **Fats** — butter, margarine, cheese, milk, fatty meat and fish, ghee.
3. **Proteins** — meat, eggs, fish, nuts, cheese, soya bean.
4. **Vitamins** — see Table 4.1. Note that only vitamin D can be manufactured by the body through the action of sunlight on the skin.

Table 4.1 Vitamins: their sources and characteristics

Vitamin	Source from normal food	Special source	Symptom of deficiency	Special notes
A Retinol	Liver, egg-yolk, green vegetables, red palm oil, cocoa, carrots	Butter, margarine, cod-liver oil	Sore eyes, reduced night vision, colds and bronchitis, unhealthy skin	Carotene from plant pigment converted to vitamin A in intestinal walls
B_1 Thiamine	Unpolished cereals, palm wine, beans, lean meat, egg-yolk	Bread, milk, kidney	Retarded growth, lack of appetite in children, nervous inflammation and weakness, paralysis, the disease called beri-beri	Likely in rice-eating peoples
B_2 Riboflavin	As for B_1, plus green vegetables	As above for B_1, plus Marmite and liver	Skin disorders, eye and mouth membrane sores, the disease called dermatitis	
Nicotinic acid	As above for B_1, plus green vegetables and yams	As above for B_1	Digestive disorders, mental disorders, the skin disease called pellagra	Likely in maize-eating peoples
C Ascorbic acid	Fresh fruit, citrus fruit (e.g. oranges, lemons, grapefruit), raw vegetables	Prepared concentrated juices	Bleeding from gums and other membranes, teeth disorders, reduced resistance to infection, the disease called scurvy	Common fatal disease on old sailing ships where sailors had no fresh fruit
D Calciferol	Liver, fat, fish, egg-yolk; formed in the skin by sunlight	Butter, margarine	Weak bones (particularly leg bones), poor teeth, the disease called rickets	Young mammals susceptible to disease, unlikely in tropics where plenty of sunshine
K	Liver, green vegetables, egg-yolk unpolished cereals		Prolonged bleeding; essential for blood clotting	Made by bacteria in the gut

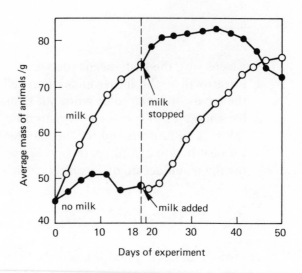

Fig. 4.5 Gowland Hopkins's experiment on feeding milk to rats

Sir Frederick Gowland Hopkins's classical experiment (see Fig. 4.5) showed how a group of rats fed on pure carbohydrates, fats, proteins, mineral salts and water hardly grew for 18 days, whereas a control group on the same diet but with 3 cm^3 of milk daily grew rapidly. After 18 days the 3 cm^3 of milk was transferred to the experimental group from the control group. Examination of the curves shows how the experimental group immediately began to increase in the rate of growth, while the control group now slowed down. Thus, the milk must have contained small quantities of chemicals (vitamins) essential to growth.

5. **Mineral salts** — see Table 4.2.

Table 4.2 Mineral elements: their sources and characteristics

Element	Source from normal food	Importance to the human body
Nitrogen N	Protein foods, lean meat, fish, eggs, milk	For synthesis of protein and other complex chemicals; formation of muscle, hair, skin and nails
Sulphur S	As for nitrogen	As for nitrogen
Phosphorus P	As for nitrogen	For synthesis of protein and other complex chemicals; formation of bones and teeth; formation of ATP
Iron Fe	Liver, green vegetables, yeast, eggs, kidney	Forms haemoglobin in red blood cells. Absence causes anaemia
Calcium Ca	Milk, cheese, green vegetables	Formation of bones and teeth; necessary for muscle contraction and blood clotting. Absence causes rickets
Iodine I	Sea fish and other sea foods, cheese, iodised table salt	Formation of hormone in the thyroid gland. Absence causes goitre and reduced growth
Sodium Na	Table salt, green vegetables	Maintenance of tissue fluids, blood and lymph; transmission of nerve impulses
Potassium K	Vegetables	Transmission of nerve impulses
Chlorine Cl	Table salt	Maintenance of tissue fluids, blood and lymph

6. **Water** — see discussion of homeostasis and excretion in Chapter 8 for the importance of a water balance.
7. **Dietary fibre (roughage)** — consists mainly of cellulose derived from plant cell walls. The diet must consist of fruit, vegetables and bread to provide roughage, but these can be supplemented by bran. Fibre adds bulk to the faeces and regular daily bowel movement becomes easier. Lack of fibrous material may cause constipation.

Table 4.3   Comparison of an internal combustion engine and Man

	Engine	*Man*
Preparation of fuel	Crude oil purified	Food digested
Fuel used	Carbon compounds of C, H and O — petrol	Carbon compounds of C, H and O — carbohydrates and fats
Supply of oxygen	From air through the carburettor	From air through the lungs
Combustion	Oxidised fuel in a chamber	Oxidised food in a cell
Waste products	Carbon dioxide, water and other gases	Carbon dioxide and water
Energy production	Heat and mechanical energy	Heat and mechanical energy
Transfer of energy	By a system of cranks and gears	By a system of muscles, bones and joints

4.4   Energy value of food

Energy is required for: (1) **maintaining living processes** such as heart beat, breathing, circulation, etc. (Basal Metabolic Rate: BMR); (2) all other activities — (a) **everyday activities** (sitting, standing, etc.) and (b) **the work we perform**.
BMR depends upon:

1. **age** — more energy for growth is required when young and less when mature and old.
2. **size** - more energy is required for small children and thin adults, less for fat adults.
3. **climate** — more energy is required for life in cold climates and less in hot climates.

Under controlled laboratory conditions:

1 g of glucose produces in the body about 16 kJ of energy;
1 g of fat produces in the body about 38 kJ of energy;
1 g of protein produces in the body about 17 kJ of energy.

Thus, fats have about 2½ times the energy value per unit mass of carbohydrates. The energy requirements for three broad groups of men are as follows:

	Sedentary	*Moderately active*	*Very active*
BMR	1 900	1 900	1 900
Everyday activity	5 000	5 000	5 000
Work (occupation)	3 400	4 600	6 800
	10 300 kJ	11 500 kJ	13 700 kJ
	= 10.3 MJ	11.5 MJ	13.7 MJ

The values for women are somewhat lower, since they need to expend comparatively less energy because of their smaller size.

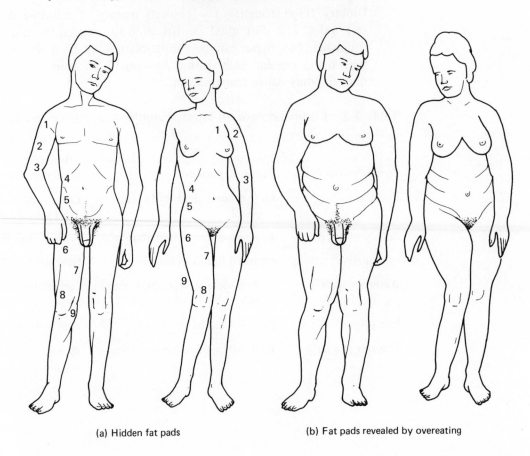

(a) Hidden fat pads   (b) Fat pads revealed by overeating

Fig. 4.6   (a) A 29-year-old man and woman and hidden fat pads 1–9; (b) excess of fat reveals position of fat pads

Table 4.4   Fat pads

Fat areas numbered in Fig. 4.6	Desirable average thickness of each fat pad (mm)	
	Men	Women
1. Shoulder	18.0	17.8
2. Outside arm	4.4	6.2
3. Inside arm	3.5	6.6
4. Hip	11.2	19.0
5. Top of thigh	15.6	28.1
6. Outside leg	4.8	7.4
7. Inside leg	6.0	10.9
8. Front of leg	7.0	13.0
9. Knee	3.0	6.5

4.5   Balanced diet

See definition on page 42. Daily food intake should be a mixed diet of:

    1. meat, eggs, fish, milk;
    2. vegetables and fruit;
    3. grain cereals, potatoes, bread.

Each of these foods provides certain essentials but may be deficient in others, e.g. vitamin C is lacking in 1 and 3 but present in 2, while vitamin D is lacking in 2 and 3 but present in 1.

In a balanced diet nutrients are eaten in the correct amounts for age and activity. The following factors must be taken into consideration.

1. The amount of energy required.
2. The minimum protein requirements, taking into account that one-fifth of the protein should be animal protein.
3. The fat content, which should supply one-sixth of the total energy requirements.
4. The carbohydrate content, which should supply five-sixths of the total energy requirement.
5. The other essential nutrients, vitamins and mineral salts, although only needed in very small quantities, are absolutely vital.
6. Sufficient fluids should be included to complement that contained in the 'solid' foods.
7. An element of roughage should be included to stimulate bowel action.

These principles also apply to vegetarians, except for point 2. In their case, the total amount of protein must come from plant sources, unless they eat such 'marginal' foods as eggs and cheese.

Table 4.5   Daily diet of woman doing light factory work

Meal	Energy (kJ)	Protein (g)	Fat (g)	Calcium (mg)	Iron (mg)	Vitamin A (i.u.)	Thiamine (mg)	Riboflavin (mg)	Ascorbic acid (mg)
Breakfast	2 982	21	26	240	4	630	0.51	0.30	5
Snack	882	9	8	140	1	230	0.12	0.12	0
Lunch	3 486	20	27	350	4	1 100	0.87	1.90	30
Snack	1 008	5	9	50	1	50	0.06	0.05	0
Evening meal	2 394	19	21	100	4	560	0.51	1.12	31
Total	10 752	74	91	880	14	2 570	2.07	3.49	66

Table 4.6   Analysis of two meals

	Weight (g)	Energy (kJ)	Protein (g)	Fat (g)	Calcium (mg)	Iron (mg)
Meal 1						
Roll	98	1 058	8.8	1.1	109	1.8
Butter	8	273	0.0	7.3	0	0.0
Cheese	56	983	14.2	19.6	460	0.4
Tea (with milk and sugar	280	84	1.0	1.0	30	0.0
Total	442	2 398	24.0	29.0	59.9	2.2
Meal 2						
Mutton	67	945	8.9	21.1	7	1.4
Cabbage	84	88	1.2	0.0	54	0.9
Potato	112	353	2.4	0.0	8	0.8
Apples	84	151	0.3	0.0	3	0.3
Custard	56	260	1.8	2.2	70	0.0
Total	403	1 797	14.6	23.3	142	3.4

4.6 Relationship between calories and joules

All energy is transferable, so that, for whatever purpose it is required, the energy value of foods and energy requirements of Man can be calculated in terms of heat energy. Heat energy for many years has been measured in calories, which are defined as follows:

1 **calorie** is the heat required to raise the temperature of 1 g of water through 1 degree C.

For practical purposes in nutrition this unit was too small, so that the **kilocalorie** was used:

1 **kilocalorie** (= 1000 calories) is the heat required to raise the temperature of 1000 g of water through 1 degree C (or 100 g of water through 10 degrees C, and so on).

With the introduction of SI units, the unit of heat used is now the **joule**, which is defined in terms of individual work done. James Joule showed that **mechanical work** can be converted into an **equivalent amount of heat**. Therefore both can be measured in the same units. For our purposes all we need to consider is that:

1 calorie (cal)	= 4.2 joules (J)
1 kilocalorie (kcal)	= 4200 joules
	= 4.2 kilojoules (kJ)

In many books on nutrition the energy value of foods is still described in terms of kilocalories, but kilojoules and megajoules are the current units:

joule	J	1000 J	= 1 kJ
kilojoule	kJ	1000 kJ	= 1 MJ
megajoule	MJ	1 MJ	= 1 000 000 J
1 MJ	= 238 kcal		
1 kcal	= 4.2 kJ		

Remember that 1 g of water has a volume of 1 cm^3; thus, 10 g of water has a volume of 10 cm^3.

You can use this knowledge to find out how much heat energy is released from food when it is burned. Carbohydrates and fats in the food are the principal sources of energy for the body, but protein can also be used for this purpose. If these substances are burned under controlled conditions in the laboratory, the energy produced can be calculated. Typical results:

1 g of carbohydrate (as glucose) produces in the body about 16 kJ;
1 g of fat produces in the body about 38 kJ;
1 g of protein produces in the body about 17 kJ.

Notice that **fats** have about $2\frac{1}{2}$ times the energy per unit mass of **carbohydrates**.

Using these values, the energy content of any food can be calculated from the proportions of nutrients that it contains. For example, 100 g of maize (whole grain) contains 10 g of protein, 4.5 g of fat and 70 g of carbohydrate. Thus, the total energy value can be worked out as shown:

$$70 \times 16 \text{ kJ} = 1120 \text{ kJ from carbohydrate}$$
$$4.5 \times 38 \text{ kJ} = 171 \text{ kJ from fat}$$
$$10 \times 17 \text{ kJ} = \underline{170 \text{ kJ from protein}}$$
$$1461 \text{ kJ}$$

4.7 Questions and answers

(a) Multiple-choice Questions

1 Which one of the following has the greatest amount of energy per unit volume?
  A cane sugar
  B corn oil
  C molasses
  D potato starch
  E lean beef

2 Which of the following has the most chemical elements present in all proteins?
  A nitrogen only
  B hydrogen and nitrogen
  C nitrogen and oxygen
  D hydrogen, nitrogen and oxygen
  E nitrogen, oxygen, hydrogen and carbon dioxide

Table 4.7 Constituents of some foods

	Protein (g)	Fat (g)	Carbohydrate (g)	Vitamin A (mg)	Energy value (kJ)
A	27	45	14	0	2 390
B	1	0.3	10	3.0	202
C	35	18	12	1.6	1 470
D	1.2	23	8	0.12	420
E	11	16	9	0.10	632

Table 4.8 Analysis of some foods

	Protein (g)	Fat (g)	Carbohydrate (g)	Vitamin A (mg)	Vitamin C (mg)	Energy value (kJ)
A	21	0	14	0.03	10	160
B	4	4	4.6	0.04	1	294
C	1.5	0.2	6	0.22	150	134
D	0.8	0	10	0.06	45	176
E	1.7	0	11.3	0	15	205

Examine Tables 4.7 and 4.8.

3 Assuming that the amounts of these foods are the same, which one of the foods in Table 4.7 would give the best value in the diet of a growing child (amounts per 100 g of food)?

4 Which one of the foods in Table 4.8 would be most suitable in a daily diet for a child suffering from a deficiency disease whose symptoms were bleeding from the gums, teeth disorders and reduced resistance to infection (amounts per 100 g of food)?

5 Which one of the following is the organic material of an enzyme?
  A carbohydrate
  B fat
  C protein
  D vitamin
  E mineral salt

6 Which one of the following is the correct term used for the results of heating a protein?
  A killed
  B denatured
  C congealed
  D dehydrated
  E digested

7 Which one of the following vitamins is manufactured in the skin in the presence of sunlight?

   A vitamin A

   B vitamin B

   C vitamin C

   D vitamin D

   E vitamin E

8 Which one of the following is the chief use of protein in the body?

   A a source of urea

   B a food reserve

   C a bone salt

   D a growth material

   E a source of energy

9 Which one of the following results from a diet consisting principally of polished rice?

   A scurvy

   B beri-beri

   C sexual deficiency

   D rickets

   E pellagra

10 Calcium is in the human diet for

   A making bones and teeth

   B the formation of thyroxine

   C producing seminal fluid

   D balancing of tissue fluids

   E formation of hydrochloric acid

11 In order to prevent constipation by providing bulk, roughage (fibre) is essential in the diet. Roughage is made up of

   A dead organisms in the gut cavity

   B plant material consisting mainly of cellulose

   C dead cells sloughed off the gut wall

   D any tough food taken into the gut

   E waste material concentrated in the rectum

12 Which one of the following is the approximate expenditure of energy by a moderately active man during a 24 hour period?

   A 6000 kJ

   B 9500 kJ

   C 10 300 kJ

   D 11 500 kJ

   E 13 700 kJ

13 Which one of the following foods produces most energy?

   A 1 g of glucose

   B 1 g of fat

   C 1 g of amino acid

   D 1 g of protein

   E 1 g of vitamin

14 In which of the following forms is digested protein absorbed through the gut wall?

   A glucose

   B glycerol

   C fatty acid

   D amino acid

   E fructose

15 Which of the following most closely describes vitamins?

   A organic compounds containing sufficient energy to satisfy the needs of an organism

   B Man-made substances required by humans

   C substances produced by a mould

   D necessary substances which Man cannot make himself

   E complex naturally occurring chemicals

16 Which one of the following, added to the diet, would most improve the growth of teeth if the child concerned lived in an industrial slum in a northern country?
    A vitamin C
    B phosphorus
    C vitamin E
    D vitamin D
    E calcium

17 The day's diet should always contain protein because
    A it supplies all the energy
    B it tastes pleasant
    C it cannot be stored in the body
    D it aids digestion of amino acids
    E it prevents the deficiency disease of rickets

18 Iodine is essential in the diet because
    A it is used to form haemoglobin
    B it prevents tooth decay
    C it is used to form thyroxine
    D it kills bacteria in the gut
    E it is used to form adrenaline

19 Which one of the following is the main constituent of fish or meat?
    A fat
    B vitamins
    C carbohydrate
    D starch
    E protein

(b) Structured Questions

1 Using each of the foods listed below once only:

apples; cabbage; cheddar cheese; liver; milk; white bread

state which food:
(a) is rich in protein and fat but contains little carbohydrate; . . . .
(b) is fairly balanced with protein, fat and carbohydrate; . . . .
(c) contains a lot of carbohydrate and some protein; . . . .
(d) is rich in protein and iron but is low in fat; . . . .
(e) is a good source of vitamin C and contains some mineral salts; . . . .
(f) contains carbohydrates and roughage; . . . .   **(OLE)**

2 Study Table 4.9 and answer the questions on it.

Table 4.9

100 g food	Energy (kilojoules)	Protein (g)	Fat (g)	Carbohydrates (g)	Vitamins
Brown bread	993	9.2	1.8	49.0	B
Butter	3122	0.5	82.5	0	AD
Cheese	1726	25.4	34.5	0	ABCD
Boiled ham	1768	16.3	39.6	0	B
Orange	147	0.8	0	8.5	ABC
Banana	318	1.1	0	19.2	ABC
Milk	272	3.3	3.8	4.8	ABCD
Coffee	0	0	0	0	0

(i) Which food has the highest energy value?   (one line)
(ii) Which food has no vitamins?   (one line)
(iii) Which food has the lowest energy value?   (one line)
(iv) Name two foods which are rich in protein.   (two lines)
(v) Which food would you choose for a two-year-old child to make sure it had plenty of vitamins?   (one line)
(vi) Carol eats 200 g of brown bread with 50 g of butter. What is the total energy value of the meal? (Show your working)   (four lines)   **(EMREB)**

3 Table 4.10 lists the approximate food values for 10 g of certain foods. Table 4.11 is a diet sheet, to be completed, showing the food values of a diet. It is a diet for one day's food for a heavy manual worker.

Table 4.10

Food (10 g)	Energy value (kJ)	Protein (g)	Fat (g)	Vitamins
Beef	100	1.2	1.5	A B_1 B_2 B_7
Dhall	140	2.0	–	B_1 B_2
Ghee (oil)	350	–	9.2	–
Rice, polished	150	0.7	0.1	–

Table 4.11

Food	Quantity (g)	Energy value (kJ)	Protein (g)	Fat (g)
Rice, polished	600			
Beef	50			
Dhall	200	2800	40	0
Ghee	25	875	–	23
Spinach	150	150	4.5	0
Pumpkin	100	100	1.0	0
Papaya (paw paw)	400	600	4.0	0
Sugar	50	800	0	0
Coffee	10	0	0	0
Cheese	50	875	12.5	17.5
Onions	50	50	0.5	0
Spices	50	0	0	0
Total				

(a) Complete Table 4.11, and add up the figures to find
   (i) the total energy value for the diet
   (ii) the total weight of protein
   (iii) the total weight of fat.
(b) (i) If 1 g of protein provides 17 kJ, say whether the protein content of the diet is too high, too low, or approximately correct. . . . .

   . . . .

   (ii) Give reasons for your answer to (i). . . . .

   . . . .

(c) Which of the foods listed in Table 4.11 provides the best source of vitamin A? (one line)
(d) Which of the foods in Table 4.11 provides the best source of vitamin B_1? (one line)

(e) Name the deficiency disease caused by a lack of
    (i) vitamin B_7 (niacin)   (one line)
    (ii) vitamin B_{12} (cobalamin)   (one line)   **(UCLES)**

4 (c) The label below was on the side of a packet of breakfast cereal (e.g. cornflakes):

NUTRIENTS PER SERVING			
Protein	5.3 g	Niacin	0.24 g
Fat	0.13 g	Thiamine (vitamin B1)	0.22 g
Carbohydrate	22.0 g	Riboflavin (vitamin B2)	0.48 g
Iron	3.2 g	Crude fibre	0.23 g

A serving of this breakfast cereal provides one-sixth of the daily requirement of Iron, Niacin, Thiamine and Riboflavin.

    (i) What is the main nutrient found in the breakfast cereal?   (one line)   **(1)**
    (ii) What is the daily iron requirement of an adult?   (one line)   **(1)**
    (iii) Most people put milk on their cereal. Name one nutrient found in milk which is not found in the cereal.   (one line)   **(1)**
    (iv) What would be the value of eating fruit such as blackcurrants with the cereal? (one line)   **(1)**
    (v) If you were trying to be slim, why might it be better to eat grapefruit for breakfast instead of cereal and milk?   (two lines)   **(1)**
    (vi) Which nutrient in the breakfast cereal provides roughage?   (one line)   **(1)**
                          [Part question] **(YHREB, 1985)**

5 Table 4.12 shows some of the daily energy and nutritional requirements of individuals in this country.

Table 4.12

	Boys 12–15 years	Girls 12–15 years	Men	Women
Energy in kJ	11 700	9600	11 600	9200
Protein in g	70	58	75	55
Calcium in mg	700	690	510	510
Iron in mg	14	14	10	12

(a) What is the mean (average) protein requirement for the four groups shown here? (one line)
(b) Which group needs most calcium?   (one line)
(c) Name one major item of our diet that is missing from the table.   (one line)
(d) Why does a woman's need for calcium increase when she is
    (i) pregnant?   (one line)
    (ii) breast feeding a baby?   (one line)
(e) Give one good source of protein.   (one line)
(f) Which type of food provides most of the energy in our diet?   (one line)

                                    (NWREB)

6 Identify each of the vitamins or elements described in Table 4.13. Enter your answer in the space provided.

Table 4.13

Use of vitamin/element	Vitamin/element
The absorption of calcium in the ileum	*Vitamin D*
The formation of haemoglobin	*Iron*
The formation of visual purple	*Vitamin A*

7 Table 4.14 shows the energy value, the protein, fat and carbohydrate content of several different foods. All the values are per gram of the food.

Table 4.14

Food	Energy value (kJ per g)	Protein (g)	Fat (g)	Carbohydrate (g)
Bacon	18.5	0.24	0.38	0
Bread	10.0	0.07	0.01	0.4
Butter	30.5	0	0.85	0
Egg	6.0	0.12	0.11	0
Marmalade	11.0	0.001	0	0.7
Milk	2.5	0.03	0.04	0.47
Orange juice	1.5	0.004	0	0.08
Sausage	15.0	0.11	0.24	0.11
Sugar	17.0	0	0	1.05
Tea	0	0	0	0
Wholegrain cereal	16.0	0.08	0.01	0.85

(a) Use the data in the table to work out the total energy value of each part of the breakfast meals below and then the total for each meal.

(i) **Breakfast A**

Boiled egg (60 g) . . . . *360 kJ*

1 slice of toast (60 g) . . . . *600 kJ*

with butter (5 g) . . . . *152.5 kJ*

and marmalade (10 g) . . . . *110.0 kJ*

Cup of tea (160 g) *0*

with milk (50 g) . . . . *125.0 kJ*

Total Energy Value = ... *1347.5 kJ* . . . . . **(2)**

(ii) **Breakfast B**

Orange juice (100 g). . . . . *150.0 kJ* . . . .

Wholegrain cereal (30 g) . *480.0 kJ* . . . .

with milk (100 g) . . . . . . *250.0 kJ* . . . .

and sugar (10 g) . . . . *170.0 kJ* . . . .

Grilled bacon (30 g). . . . . *555.0 kJ* . . . .

and sausage (50 g) . . . . . . *750.0 kJ* . . . .

One slice of toast (60 g) . . *600.0 kJ* . . . .

Total Energy Value = . . . *2955 kJ* . . .   (2)

(b) Give **two** reasons why Breakfast **B** is a more balanced meal than Breakfast A.

(i) . . . *Provides a greater amount of protein (bacon)* . . . .

for growth. . . . .

(ii) . . . *Provides a greater amount of fibre (cereal)* . . . .

for roughage. . . . . (2)

(c) (i) Which **one** of the foods shown in the table would be the most suitable for the growth and repair of body cells?

Bacon . . . .

(ii) Explain your answer.

. . . *This food has the greater amount of* . . . .

. . . *protein, which is used for growth and repair.* . . . . . . (1)

(d) Describe the part played by each of the following in the human body:
(i) Fat

. . . *Used as a storage substance and when* . . . .

. . . *released can be respired to provide energy* . . . . (1)

(ii) Carbohydrate

. . . *Simple carbohydrate (glucose) used to provide energy.* .

. . . *Glycogen, a complex carbohydrate, is stored.* . . . . (1)

(e) Name **three** other classes of food, in addition to carbohydrate, fat and protein, which are essential for a balanced diet.

. . . . . . *Vitamins, minerals, water* . . . . (2)

(f) (i) Describe how you would test for the presence of protein in an egg.

. . . *Place a raw egg in a test-tube and* . . . .

. . . *add several drops of biuret reagent.* . . . . (1)

(ii) State the likely result which would be obtained.

. . . *A violet colour* . . . . (1)

(UCLES)

(a) What is a vitamin? **(3)**

(b) What precautions would you take to preserve the vitamin C when cooking vegetables?
(3)

(c) Why must a pregnant woman have vitamin D in her diet? **(2)**

(d) How do calcium ions from the food of a pregnant woman reach the bones of the foetus inside her? **(12)**

(UCLES)

Answer

(a) A vitamin is an organic chemical required in minute quantities by the body in order to perform essential chemical activities.

(b) Vegetables should be cooked in water. No additional substance should be added to the water, such as table salt or bicarbonate of soda, which destroys the vitamin C. The vegetables should not be soaked before cooking and the quantity of water used must be the minimal necessary to cover the vegetables. The water must be brought to the boil before the vegetables are added. Cooking should not be prolonged.

(c) Vitamin D is important for bone formation. Although vitamin D will form in the skin of the woman when exposed to sunlight, this will not be sufficient. The diet must therefore be supplemented with vitamin D to aid bone formation in the foetus.

(d) Calcium ions are present in vegetables, with large amounts in potatoes, sweet corn and fruit. Cereal grains also contain calcium ions, so that bread and biscuits are also valuable foods.

When digestion has been completed in the gut, the calcium ions are liberated and enter the blood circulation at the capillaries in the villi of the small intestine.

The blood is collected from the vessels of the gut and passes by way of the hepatic portal vein to the liver. After passing through the liver capillaries, the blood returns to the venous system by the hepatic vein. The blood finally enters the right atrium by the venae cavae. The contraction of the heart muscles (systole) forces the blood from the right atrium to the right ventricle and the ventricular contraction pushes the blood into the pulmonary circulation. It passes by way of the pulmonary artery, the lung capillaries and the pulmonary vein back to the left atrium of the heart. Systole then empties the blood into the left ventricle and its contraction forces the blood into the aorta.

The blood eventually reaches the vessels and blood spaces in the wall of the uterus. These run alongside the vessels in the placenta, so that the calcium ions are passed through the vessel walls into the foetal circulation. The blood passes into the foetus through the umbilical vein. The blood travels through the heart and into the arterial system of the foetus, eventually reaching the growing bones.

The demands of the growing foetus are so great that if the mother's diet is not supplemented, calcium ions may be withdrawn from the bones of her body.

Notes

1. This question demands a straightforward written answer and no diagrams would really be appropriate here.

2. Sections (a), (b) and (c) demand factual recall of these few important points. Section (d), however, carries twelve marks and careful consideration must be

given to exactly what is required by the examiner. It is clear that this section is more about digestion and blood circulation than diet. Nevertheless it is important to establish first the origin of the calcium ions in the diet.

3. The bulk of section (d), however, must be an account of the pathway by which the ions travel from the blood capillaries of the villi to the growing bones of the foetus.

(c) Answers to Objective and Structured Questions

(i) *Multiple-choice Questions*
1. B 2. D 3. C 4. C 5. C 6. B 7. D 8. D 9. B 10. A 11. B 12. D
13. B 14. D 15. D 16. D 17. C 18. C 19. E

(ii) *Structured Questions*

1 (a) Cheddar cheese (b) Milk (c) White bread (d) Liver (e) Cabbage (f) Apple

2 (i) Butter (ii) Coffee (iii) Coffee (iv) Cheese, boiled ham (v) Milk
   (vi) Bread 993×2 kJ $= 1986$ kJ
        Butter $3122 \times \frac{1}{2}$ kJ $= 1561$ kJ
        Total = 3547 kJ

3 (a)   *energy value*   *protein*   *fat*
        Rice $60 \times 150 = 9000$ kJ $60 \times 0.7 = 42$ g $60 \times 0.1 = 6$ g
        Beef $5 \times 100 \;\; = 500$ kJ $5 \times 1.2 \;\; = 6$ g $5 \times 1.5 = 7.5$ g
        (i) $9000 + 500 + 6250 = 15\,750$ kJ
        (ii) $42 + 6 + 62.5 = 110.5$ g
        (iii) $6 + 7.5 + 40.5 = 54.0$ g
   (b) (i) Energy value not relevant to consideration of protein content of diet.
       (ii) Protein is not used for energy except in cases of extreme starvation. Protein is only used for growth and repair of tissues.
   (c) Beef (d) Dhall (e) (i) Pellagra, (ii) Anaemia

4 (c) (i) Carbohydrate (ii) $3.2 \times 6 = 19.2$ (iii) Water/calcium
      (iv) Provide vitamin C (ascorbic acid)
      (v) This would lessen the intake of carbohydrate and fat. Fat not present and carbohydrate only as small amounts of sugar in grapefruit.
      (vi) Crude fibre 0.23 g

5 (a) $70 + 58 + 75 + 55 = 258$ divided by 4 = 64.5 g
   (b) Boys 12–15 years
   (c) Vitamins or roughage or water
   (d) (i) To maintain calcium levels in her bones and for the foetal bones.
       (ii) To provide calcium for the growth of the baby's bones.
   (e) Lean meat or fish
   (f) Fat or oil

Questions 6 and 7 have the answers supplied with the questions.

5 Digestion and Absorption

5.1 Digestion

Digestion is the breakdown of large, insoluble molecules into smaller, soluble, diffusible molecules to enable them to pass through the wall of the alimentary canal. There are two types of digestion: (1) **mechanical** – breaking down food into smaller pieces by teeth and the churning action of the gut; (2) **chemical** – breaking down food by the action of organic catalysts (enzymes) which speed up the reaction.

(a) Mechanical Digestion

Mechanical digestion begins with the teeth in the buccal cavity. There are four types of teeth: **incisors** (i) and **canines** (c) in the front of the jaws, and **premolars** (p) and **molars** (m) behind the canines in the upper and lower jaws (see Fig. 5.1).

Man as a mammal has two sets of teeth in a lifetime: the first or **milk set** consists of incisors, canines and premolars; the second or **permanent set** consists of all four types of teeth.

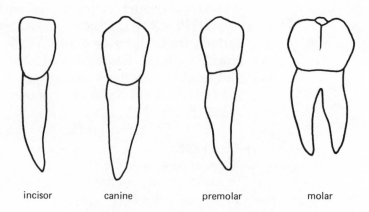

incisor   canine   premolar   molar

Fig. 5.1 Four types of teeth in Man

(i) *Dental Formula*

Since all jaws are symmetrical, only one half of each jaw is represented in a dental formula. The four types of teeth are indicated by their initial letter in each case. Thus, for Man the dental formula is

$$i \frac{2}{2} \quad g \frac{1}{1} \quad p \frac{2}{2} \quad m \frac{3}{3}$$

The total number of teeth in the permanent set is $2\left(\frac{2}{2} \quad \frac{1}{1} \quad \frac{2}{2} \quad \frac{3}{3}\right) = 32$

The milk set has only 20 teeth, and these are replaced about the age of 7 years by the permanent set.

For the structure of a tooth see the figure in structured question **1**.

(ii) *Functions of Parts of the Tooth*

1. **Enamel**: outer layer, non-living, very hard — protects underlying layers.
2. **Dentine**: living, softer than enamel, similar to bone — forms the bulk of the tooth.
3. **Pulp cavity**: filled with connective tissue, capillaries and nerve fibres (pulp). Blood supplies food and oxygen to the living cells of the dentine.
4. **Cement**: bone-like; attaches the tooth to its socket.
5. **Fibres**: absorb shock of hard bites (protect against damage).

(iii) *Dental Care*

Teeth must be used in the correct way, must be kept clean and must receive regular expert attention.

1. Teeth should never be used for an improper purpose such as cracking nuts, opening crown bottle tops, etc. A balanced diet must include food to give the teeth hard work, e.g. apples, nuts, oranges, bran, etc.
2. Teeth must be cleaned regularly. Bacteria grow on food residues between teeth and produce acids which dissolve away enamel and dentine. Toothpicks and toothbrushes should be used after a meal together with an alkaline toothpaste to neutralise acids.
3. Teeth must be examined at least twice a year by a dentist to discover any signs of decay and then to control it by filling cavities.

(iv) *Dental Disease*

There are two classes of disease:

1. **Dental caries**, which is caused by (a) lack of hard food; (b) too much sweet food; (c) lack of calcium in the diet; (d) lack of vitamin D; (e) lack of cleaning; (f) general ill-health.
2. **Periodontal disease**, which is caused by (a) lack of vitamins A and C; (b) lack of gum massage (by toothbrush); (c) imperfect cleaning.

As a result of 1 and 2 above, **plaque** may collect. Plaque is a soft sticky paste formed from saliva and food, and is a site of bacterial activity. The poisons produced by the bacteria may penetrate the gum and cause inflammation. This disease is called pyorrhoea.

(v) *Fluoridation*

Fluoride salts in drinking-water, in toothpaste or in tablet form help in preventing dental caries (see Fig. 5.2). Scientific opinion is divided about the practice of adding fluoride to drinking-water. In areas in which fluoride is present naturally in some drinking-water there is a clear improvement in the incidence of dental caries. In highly industrialised countries, however, other sources of fluoride may boost the amount received by the individual; thus, overdosage is possible.

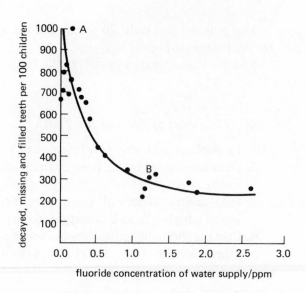

Fig. 5.2 Graph showing the effects of fluoride concentration on tooth decay in children

(b) Chemical Digestion

The process of chemical digestion begins in the buccal cavity. Chewing of food results in the production of **saliva** containing mucus, salts and **salivary amylase** (ptyalin). This enzyme breaks down starch into **maltose** (sugar). Saliva also contains **antibodies** that help to destroy bacteria.

While being chewed by the premolars and molars, the food is acted upon by saliva, which dissolves, digests and moistens it. The particles of food are rolled into a ball (**bolus**) by the tongue. The bolus is pushed to the back of the buccal cavity, where the swallowing reflex occurs. At this point the opening (**glottis**) to the larynx and trachea is closed by a flap, the **epiglottis**, to prevent blockage of the airway. The bolus enters the oesophagus and is pushed down by **peristalsis** (see below and Fig. 5.4).

For the **structure of the alimentary canal**, see Fig. 5.3.
For the **action of enzymes** in the alimentary canal, see Table 5.1.
For the **action of muscles** pushing food along the gut, see Fig. 5.4.

5.2 The alimentary canal

1. **Digestion** – see previous definition and Table 5.1. The process begins in the buccal cavity, and continues in the stomach and the small intestine.
2. **Peristalsis** – waves of contraction that pass along the involuntary muscles of the canal. The action mixes the contents and pushes along the digesting food materials to the next section of the intestines. It is an action of antagonistic muscles, the circular and the longitudinal.
3. **Absorption** – occurs in the duodenum and the ileum (small intestine). Glucose and amino acids enter the capillaries of the villus by the process of diffusion, and similarly the fatty acids and glycerol enter the lacteal (part of the lymphatic system) of the villus. Folds of the intestinal wall and the villi present a greatly increased surface area, to speed absorption of digested food. Intestinal juices (succus entericus) containing enzymes are released from glands between the villi.

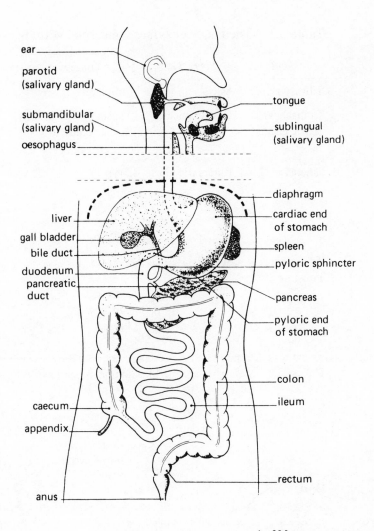

ear

parotid
(salivary gland)

submandibular
(salivary gland)

oesophagus

liver

gall bladder

bile duct

duodenum
pancreatic
duct

caecum

appendix

anus

tongue

sublingual
(salivary gland)

diaphragm

cardiac end
of stomach

spleen

pyloric sphincter

pancreas

pyloric end
of stomach

colon

ileum

rectum

Fig. 5.3   The alimentary canal of Man

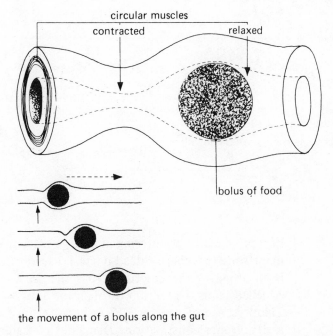

circular muscles

contracted   relaxed

bolus of food

the movement of a bolus along the gut

Fig. 5.4   Peristalsis and how food is moved along the oesophagus and intestines

Table 5.1 The major enzymes concerned with digestion

Gland	Secretion	Enzymes	Action substrate	Products
Salivary glands	Saliva	Amylase	Starch	Maltose
Stomach wall — gastric glands	Gastric juice	Pepsin + hydrochloric acid + renin	Protein Soluble milk protein	Polypeptides Insoluble milk protein
Pancrease	Pancreatic juice	Trypsin	Protein	Peptides and *amino acids*
		Lipase	Fats	*Fatty acids* and *glycerol*
		Amylase	Starch	Maltose
Wall of small intestine	Succus entericus	Sucrase	Sucrose	*Glucose* and *fructose*
		Maltase	Maltose	*Glucose*
		Peptidase	Polypeptides	*Amino acids*
		Lipase	Fats	*Fatty acids* and *glycerol*

The substances in italics are the final soluble breakdown products of digestion.

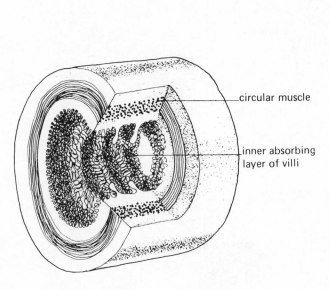

Fig. 5.5 Portion of the small intestine, showing internal folds and villi

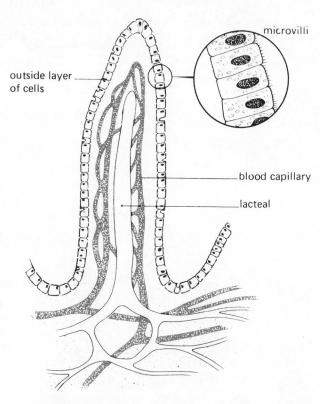

Fig. 5.6 A villus and its blood supply

4. **Faeces** — material entering the large intestine (colon) is mainly undigested plant fibre (roughage), dead bacteria, dead cells and water. Most of the water is absorbed in the colon, so that the faeces become semi-solid. These are expelled from the rectum through the anus (defaecation) at regular intervals. The presence of roughage gives bulk to the faeces, and this aids elimination and prevents constipation. Diarrhoea — watery faeces — is due to the non-absorption of water in the colon, caused by unusual diets or bacterial infection.

5.3 Transport of digested food

Digested, soluble food enters the bloodstream and is conveyed to the liver, where it undergoes change or is passed directly through and carried to the rest of the body.

5.4 Functions of the liver

See Fig. 5.7.

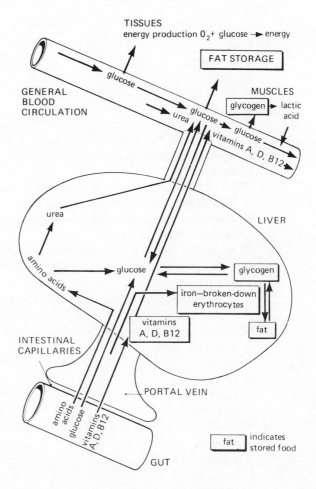

Fig. 5.7   Relationship between the liver as a storage organ and the blood circulation

1. The liver **removes glucose** from the blood and converts it to **glycogen**. If blood sugar levels fall, glycogen is converted back to glucose.
2. The liver contains about 4% lipid. **Lipids** conveyed to the liver are changed into **glycogen and glucose**.
3. **Excess amino acids** cannot be stored. They are broken down in the liver to **carbohydrate and urea (deamination).**
4. **Vitamins A and D** are stored in the liver.
5. **Vitamin B$_{12}$** is found in the liver. This is the antianaemic factor.
6. The liver **stores iron** from broken-down red blood cells, as well as glycogen.
7. The liver breaks down poisonous substances, e.g. alcohol and drugs. This process is called **detoxication**.
8. The liver produces **large quantities of heat** which warms the blood as it passes through.

5.5 Questions and answers

(a) Multiple-choice Questions

1 Which of the following is **not** a true statement about enzymes?
A They are catalysts.
B They are produced by ductless glands.
C They increase the rate of a chemical change.
D They act in only one type of reaction.
E They are still present at the end of the reaction.

2 Which of the following is concerned with the digestion of protein?
A secretin
B insulin
C pepsin
D thyroxine
E rickets

3 Which of the following enzymes is found in pancreatic juice?
A ptyalin
B pepsin
C lipase
D erepsin
E maltase

4 Which of the following is the best definition of an enzyme?
A A complex substance which can only act in the body at a particular temperature.
B A complex substance which must be present in the diet of animals in order that they remain healthy.
C A complex organic compound which accelerates the growth rate in plants.
D A complex organic compound which will only digest one particular type of food constituent.
E A complex organic compound which speeds up a particular reaction in the bodies of animals and plants.

5 Which one of the following digestive juices does not contain enzymes?
A pancreatic juice
B saliva
C bile juice
D gastric juice
E intestinal juice

6 Which of the following classes of food is acted upon by the gastric juices of the stomach?
A vitamins
B carbohydrates
C proteins
D minerals
E lipids

7 Which one of the following emulsifies lipids in order to aid their digestion?
A rennin
B mucus
C pepsin
D hydrochloric acid
E bile salts

8 Which one of the following secretions is **unable** to digest carbohydrates?
A pancreatic juice
B gastric juice
C saliva
D intestinal juice
E succus entericus

9 Which one of the following is the final breakdown product of carbohydrate?
A glycerol
B glycol

C glucose

D glycerine

E sucrose

10 Which of the following statements is true of a man with a full set of teeth?

A He has more canines than incisors.

B He has more canines than premolars.

C He has more premolars than molars.

D He has more incisors than molars.

E He has more molars than incisors.

11 Which of the following parts of a tooth protects it from bacterial attack?

A enamel

B dentine

C gum

D cement

E pulp

12 Examine Fig. 5.2. Which one of the following conclusions **cannot** be drawn from the graph?

A The decay rate does not change when the fluoride level reaches 2.0 ppm.

B The number of decayed teeth steadily decreases as the fluoride concentration rises.

C The decay rate averages 750 per hundred children between 0 and 0.5 ppm.

D Tooth decay will disappear when the concentration of fluoride is steadily increased.

E Increasing fluoride concentration helps to decrease the number of decayed, missing and filled teeth.

13 Which of the following will result from a diet without roughage?

A indigestion

B diarrhoea

C cramps

D constipation

E appendicitis

14 Which one of the following is the best definition of peristalsis?

A The reflex action of swallowing when food enters the back of the throat.

B The movement of food in a series of jerks along the gut.

C The contraction of the muscles of the colon to eliminate undigested food.

D The wave-like contraction of gut muscles pushing food along the gut.

E The contraction of oesophageal muscles driving food down to the stomach.

15 In which one of the following gut sections is a villus found?

A liver

B stomach

C small intestine

D rectum

E large intestine

16 Which one of the following is a function of the villi?

A To increase the surface area of the gut wall.

B To secrete the intestinal juices for digestion.

C To move undigested food along the gut.

D To act as a sieve for undigested foods.

E To protect the inner wall from sharp pieces of food.

17 Which one of the following products of digestion is absorbed into the lacteal of each villus?

A glucose

B amino acids

C cellulose

D fatty acids

E hydrochloric acid

18 Which one of the following produces gastric juice?

A stomach

B liver

C pancreas

      D ileum

      E colon

19 Which one of the following is the process of peristalsis?

      A It is the cause of appendicitis.

      B It allows the food to leave the stomach.

      C It stops the food going down the trachea.

      D It squeezes food along the intestine.

      E It pushes out the faeces.

20 Which one of the following is part of the process of digestion?

      A Changes food into faeces.

      B Forms food into a bolus.

      C Absorption of food through villi.

      D Changes food into a soluble form.

      E Breaks down cellulose.

(b) Structured Questions

1 The figure below shows a vertical section through a tooth.

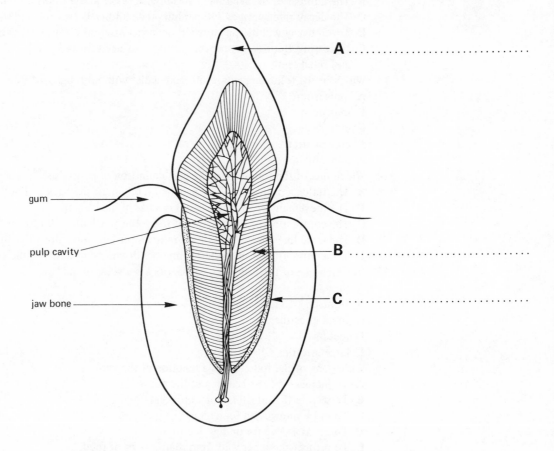

(a) Label, on the diagram, the parts marked **A**, **B** and **C**. (3)

(b) Which type of teeth present in an adult are missing from the milk teeth? (one line)
(1)

(c) Which substance, sometimes present in drinking-water, increases the resistance of the teeth to decay? (one line) (1)

(UCLES)

2 (b) In what part of the body do enzymes begin to digest fat? (one line) (1)

    (c) Name two substances that act on fat when digestion commences. (two lines) (2)

    (d) Describe the action of these two substances.

       (i) (two lines)

       (ii) (two lines) (2)

(e) Name the liquid which transports the products of digested fat away from the small intestine. (one line) **(1)**

[Part question] **(UCLES)**

3 The diagram shows a section through the stomach region of the alimentary canal of Man.

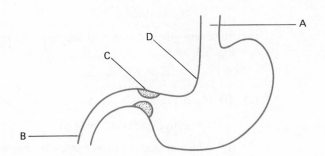

(a) Name the parts labelled **A, B** and **C** (one line each) **(3)**
(b) Name one enzyme secreted from the wall of the stomach at **D**, that would be present in the gastric juice. (one line) **(1)**
(c) (i) Name one chemical, other than an enzyme, that would be present in the gastric juice. (one line) **(1)**
    (ii) State two functions of this chemical. (two lines) **(2)**

4 The figure below shows part of the small intestine of Man. Examine the figure and answer questions (a) to (c).

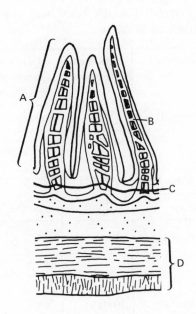

(a) Name the structures **A** to **D**.

A . . . . **B** . . . .

C . . . . **D** . . . .

(4)

(b) Describe briefly the part played by structures **B** and **C** in the overall function of **A**. (four lines) **(3)**
(c) What is the function of the structures labelled **D**? (two lines) **(3)**

5 Examine Fig. 5.5 and answer the following questions.
  (a) List two features shown in Fig. 5.5 that increases the internal surface area of the small intestine.

  (i) ..... *Villi* .....

  (ii) ..... *Microvilli* .....

  (b) State one other feature that increases the internal surface area, but is not shown in Fig. 5.5.

..... *Folded inner wall* .....

  (c) (i) Name the muscle present in the gut wall, other than the one labelled in Fig. 5.5.

..... *Longitudinal muscle* .....

   (ii) Which of these muscles contracts when peristalsis forces food along the intestine?

..... *Circular muscle* .....

  (d) Name the organic compound to which each of the following foods is converted by digestion, before absorption into the intestinal wall.

   (i) protein ..... *Amino acid* .....

   (ii) fat ..... *Glycerol and fatty acids* .....

   (iii) starch ..... *Glucose* .....

   (iv) vitamin C ..... *Vitamin C* .....

6 Complete the table below, which summarises aspects of digestion of certain nutrients and their use in the body.

Nutrient	Region of digestion	Gland(s) producing digestive juice	Digestive enzyme	Products of digestion	Use in body
Starch	*Buccal cavity*	Salivary	*Amylase*	Maltose	*After digestion to glucose; used for energy production*
	Small intestine	*Pancreas*	*Amylase*	*Maltose*	
Vitamin D	*None*	*None*	None	None	Calcification
Fats	*Small intestine*	Pancreas	*Lipase*	*Glycerol and fatty acids*	Energy
Proteins	*Stomach*	*Gastric glands*	Pepsin	*Polypeptides*	*Amino acids are used for growth and repair*
	Small intestine	*Pancreas*	Trypsin	*Peptides and amino acids*	
Sodium chloride	None	None	None	*Sodium and chlorine ions*	*Maintenance of tissue fluids, blood and lymph*

(L)
(12)

7 The graph on page 71 shows the effect of fluoride concentration on tooth decay in children. Each dot represents a different town.

Examine the graph and answer the following questions.
  (a) What is the average number of decayed, missing and filled teeth **per child** in town *A*?

..... *10* .....

70

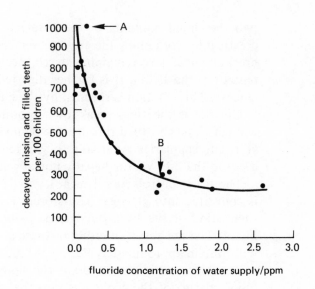

fluoride concentration of water supply/ppm

(b) What is the average number of decayed, missing and filled teeth **per child** in town *B*?

......3..

(c) 1.3 ppm is regarded as the best concentration of fluoride in public drinking water. In what way does the graph support this view?

The graph decreases steeply between 0.0 and 1.3 ppm of

fluoride, whereas between 1.3 and 2.6 ppm it levels off.

Therefore, greater amounts of fluoride in the drinking

water have no effect on tooth decay.

(d) Give **two** reasons why some local authorities oppose fluoride being added to public drinking-water.

(i) *It is a form of compulsory medication - contrary to*

the freedom of the individual.

(ii) *In highly industrialised countries pollution may*

add considerably to dosage received by water.

(UCLES)

Free-response Questions

1 Describe the changes undergone by
 (a) a starch molecule from the time it is ingested by a human until its products are broken down to provide energy; **(10)**
 (b) a protein molecule from the time it is swallowed until its structural molecules are used in a cell of the skin. **(10)**

Answer

(a) Starch, a polysaccharide, is ingested into the buccal cavity. Here it is masticated by the teeth and moved around by the tongue while at the same time being subjected to the action of an enzyme called salivary amylase, which catalyses its breakdown into a disaccharide called maltose. This is passed into the stomach by peristaltic activity of the oesophagus. Here the acidity of the stomach stops any further action of the salivary amylase. After an hour or two the liquid contents of the stomach pass into the duodenum. Here pancreatic juice containing the enzyme amylase enters the duodenum and aids the breakdown of any remaining starch into more maltose. When the maltose is passed to the ileum, it is further digested into two monosaccharide units of glucose. This reaction is speeded by the enzyme maltase.

Glucose is the final breakdown product of starch. It is soluble and small enough to be absorbed through the ileum wall into the blood capillaries of the villi. The capillaries eventually join up and the blood carries the digested food away to the liver via the hepatic portal vein.

The liver controls blood sugar levels. If there is an excess of glucose, then it is converted into glycogen and stored. If there is a deficiency of blood sugar, then some of the liver glycogen is reconverted to glucose and added to the blood. The net result of either of these activities is that blood leaving the liver will contain approximately 95 mg of blood sugar per 100 ml of blood.

The blood now passes round the body and glucose is absorbed from the blood plasma of the capillaries into the cells of different organs of the body. Within the cells the sugar is either oxidised to liberate energy, carbon dioxide and water by the process of aerobic respiration, or stored in certain cells as glycogen for use at a later date.

(b) Food containing protein is chewed and swallowed but no digestive action occurs before the food reaches the stomach. Here, in a pH of 2.00, the enzyme pepsin catalyses the conversion of protein to polypeptides. The contents of the stomach pass into the duodenum at intervals as spurts of liquid through the pyloric sphincter. Here the enzyme trypsin, contained in pancreatic juice, aids the conversion of the polypeptides into peptides and some amino acids.

Finally, when the partly digested food reaches the ileum, any remaining polypeptides and peptides are broken down into amino acids under the influence of peptidases (erepsin).

The amino acids, being soluble, are absorbed into the bloodstream via the villi of the ileum wall and pass to the liver in the blood of the hepatic portal vein. Any surplus amino acids will be deaminated in the liver, but those remaining will pass from the liver to the heart and from there some will travel via the dorsal aorta to the skin cells. The amino acids are absorbed into these cells and under the direction of nuclear DNA are linked up together to form a protein for either metabolic or structural use.

Notes

1. (a) This could be a very extensive answer and therefore each stage of the process must be dealt with briefly. It must involve the liver, which controls blood sugar levels.
2. (b) This also could be a very long answer and brevity is essential. Again the liver must be included, since this organ controls the levels and types of amino acids in the skin.
3. The quantity of material which can be written down in the time available is limited and therefore one must consider whether diagrams would save time. It would not appear so in this question, except possibly a diagram of the liver and its relations with parts (a) and (b) of the question (see Fig. 5.7).

2 (a) Make a large, clear drawing of a vertical section through a molar tooth. Label your drawing. (8)
   (b) Write a short note about the function of each of the parts you have labelled. (6)
   (c) Explain how molar and premolar teeth help to prepare food for digestion. (3)
   (d) How do bacteria cause tooth decay? (3)

(UCLES)

Answer

(a)

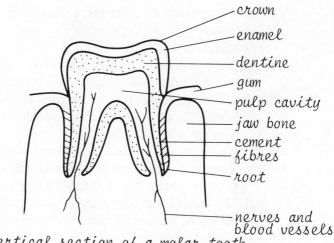

Vertical section of a molar tooth

(b) crown – for chewing and crushing food

enamel – protective, hard, dead, surface

dentine – forms the main body of the tooth

gum – protects the base of the crown; produces enamel

pulp cavity – contains pulp, living cells, nerves and blood vessels

jaw bone – provides the socket in which the tooth is embedded

cement – holds the tooth to the bone

fibres – cushion the tooth against shock

root – anchors the tooth in the jaw bone

blood vessels – bring nutrients to the tooth

nerves – provide means of cell control

(c) The premolars and molars have flat crowns which enable the teeth, together with the sideways movement of the lower jaw, to crush and chew food. This is the beginning of mechanical digestion – that is, the breaking up of food to increase its surface area. The action of enzymes is thus aided by this enlarged area on which they can act.

(d) Food remains between the teeth provide a site for the growth of bacteria. This occurs with sugary foods and particularly with sweets such as toffees. The bacteria produce acids as a result of their living processes and these acids can dissolve the enamel and dentine. The holes become points of entry for food particles and bacteria, increasing the rate of decay.

Notes

1. The drawing must be kept simple, with a minimum of shading. It should be large and clearly labelled.
2. In section (b) a 'short note' only is required as shown in the answer. Do not waste time with long and complicated sentences.
3. Sections (c) and (d) require short precise answers, each gaining three marks only. They are underlined in these two sections.

(c) **Answers to Objective and Structured Questions**

(i) *Multiple-choice Questions*

1. B   2. C   3. C   4. E   5. C   6. C   7. E   8. B   9. C   10. E   11. A   12. D
13. D   14. D   15. C   16. A   17. D   18. A   19. D   20. D

(ii) *Structured Questions*

1 (a) **A** – enamel; **B** – dentine; **C** – cement

(b) Molar

(c) Fluorine

2 (b) Small intestine   (c) Bile and lipase   (d) (i) Emulsify fats   (ii) Break down fats to fatty acids and glycerol   (e) lymph

3 (a) **A** – oesophagus; **B** – duodenum; **C** – sphincter muscle

(b) Pepsin   (c) (i) Hydrochloric acid   (ii) Kill bacteria; provide the correct medium for enzyme action

4 (a) **A** – villus; **B** – lacteal; **C** – blood vessel/capillary; **D** – muscles

(b) **B** – absorbs digested fats as fatty acids and glycerol

C – transports digested food, i.e. glucose and amino acids after absorption

(c) The function of the muscles is to push food along the intestine, and this is achieved by a rhythmic contraction of circular and longitudinal muscles called peristalsis.

Questions **5–7** have the answers supplied with the questions.

6 Transport

The transport system depends on the mass flow of a fluid, the blood, which is reviewed below.

6.1 The blood

For the components of blood and their functions, see Fig. 6.1 and Table 6.1.

Table 6.1 Summary of functions of the blood

Blood component	Functions
Red blood cells	Transport oxygen as oxyhaemoglobin. Transport carbon dioxide (very small amount)
White blood cells	Attack and engulf bacteria (polymorphs). Produce antibodies (lymphocytes)
Blood platelets	Aid clotting (together with plasma components)
Plasma	Transports:
	1. Carbon dioxide as bicarbonate from tissues to lungs
	2. Waste matter from tissues to excretory organs
	3. Hormones from ductless glands to the tissues where they act
	4. Digested food from small intestine to tissues
	5. Heat from muscles and liver to all parts of the body
	6. Ions and water to maintain the balance of body fluids
	7. White blood cells and antibodies to sites of infection
	8. Platelets and serum proteins

(a) Blood Groups

The various blood groups and their compatibility are shown in Table 6.2. In 1900 Karl Landsteiner discovered the properties of different blood groups which enabled transfusions to become more effective. He established that there were four types of blood group based on the cell factor in red blood cells:

type O – no cell factor;
type A – A-type factor;
type B – B-type factor;
type AB – AB-type factor.

Cells of one type only are present in the blood of an individual, which also contains anti-cell factors (antibody) according to the following pattern:

type O blood has anti-A and anti-B antibodies;
type A blood has anti-B antibodies;
type B blood has anti-A antibodies;
type AB has no plasma antibodies.

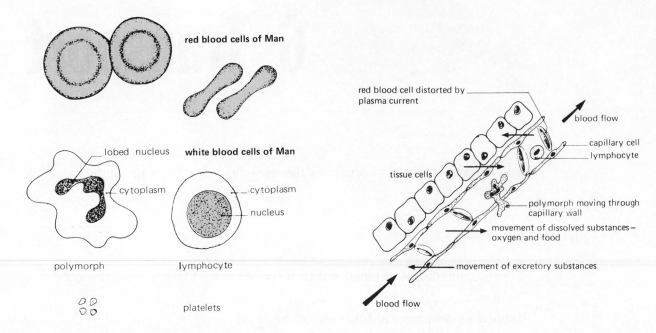

Fig. 6.1 The different types of mammalian blood cell   **Fig. 6.2** The movement of blood cells through a capillary

For transfusion purposes we must consider only the effect that the **antibodies in the recipient's** plasma will have on the **red blood cells of the donor** (see Table 6.2).

Table 6.2 Blood group compatibilities

Patient's (recipient's) blood group	Antibody present in serum	Donor's blood group			
		O	A	B	AB
O	ab	√	x	x	x
A	b	√	√	x	x
B	a	√	x	√	x
AB	o	√	√	√	√

Group AB can receive all other groups. Group AB called **universal recipient**

Group O can be given to all other groups. Group O called **universal donor**

√ = compatible with recipient.
x = incompatible with recipient, i.e. agglutinated.

(b) Rhesus Factor

Eighty-five per cent of the population have the rhesus factor antigen in the blood and are said to be Rh+; the 15% lacking the antigen are said to be Rh−. If Rh+ blood is introduced into Rh− blood (by transfusion or by 'leakage' through the uterus into the foetus), then antibodies to Rh+ factor are formed. Problems arise where an Rh− mother carries the child of an Rh+ father (the child will be Rh+). The mother develops rhesus antibodies through 'leakage' of foetal blood into her circulation. They do not affect the foetus at this pregnancy but could seep back into the bloodstream of a foetus in a second pregnancy. The red blood cells of the foetus are progressively destroyed by the antibodies thus formed (anti-Rh+), and when the child is born it suffers from haemolytic disease. The child can be saved by a massive blood transfusion shortly after birth.

(c) Blood Clotting

At a wound, where damage to blood vessels occurs, the following sequence takes place:

1. **Platelets** release an enzyme, **thrombokinase.**
2. This enzyme converts **prothrombin** into **thrombin.**
3. Thrombin converts **fibrinogen** into **fibrin** in the presence of **calcium ions** in the blood.
4. Fibrin forms a mesh of fibres which, together with trapped red blood cells, develop **a clot**. This stops blood flow.

(d) Infection

Lymphocytes produce antibodies as a result of the presence of antigens — foreign substances which invade the body, e.g. bacteria, chemicals, protein, cells and tissues. The antibodies destroy or neutralise the 'invaders'. Antibodies are either: (a) **agglutinins**, causing bacteria to clump together and then be destroyed by polymorphs, or (b) **antitoxins**, which neutralise metabolic products (toxins) of bacteria.

For immunity and vaccination, see Chapter 14.

(e) Red Blood Cells

Red blood cells have a short life of about 120 days: thus, $\frac{1}{120}$ of the circulating cells is broken down every day, and at the same time $\frac{1}{120}$ (about 25 g) of the cells is re-formed. The haemoglobin of the broken-down cells is excreted through the liver as two green pigments, **bilirubin** and **biliverdin**, in the bile.

As oxygen and air pressure decreases as one ascends in the atmosphere, people who live at high altitudes in mountains tend to produce more red blood cells (see Fig. 6.3). Notice how the red blood cell count rises as the climbers become accustomed to altitude.

Anaemia results from insufficient production of red blood cells. Patients suffer from fatigue, shortness of breath, headache and pallor. The disease may be due to lack of vitamin B_{12}, of iron or of vital proteins. **Haemolytic anaemias** are due to the destruction of red blood cells (see Section 6.1(b)). **Pernicious anaemia** is caused by the absence of vitamin B_{12}. **Sickle cell anaemia** is due to an abnormal form of haemoglobin (see Chapter 13).

6.2 The circulatory system

For the structure of the heart and the circulatory system, see Figs. 6.4 and 6.5.

The essentials of a transport system consist of a pump (the heart) and a connecting series of vessels through which blood is pumped. The blood circulates in one direction, aided by valves in the pump and vessels.

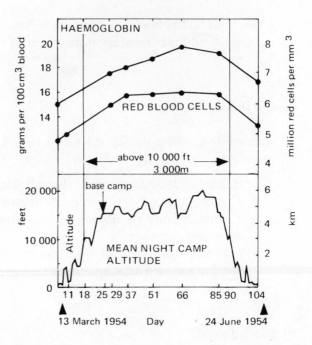

Fig. 6.3 Mean red blood cell count and blood haemoglobin level, and mean night camp altitude, of ten climbers during an attempt in 1954 to reach the summit of Mount Makalu in Nepal

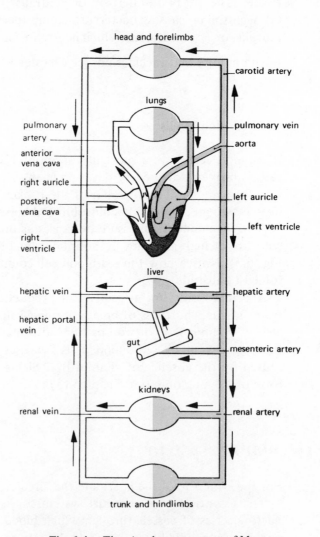

Fig. 6.4 The circulatory system of Man

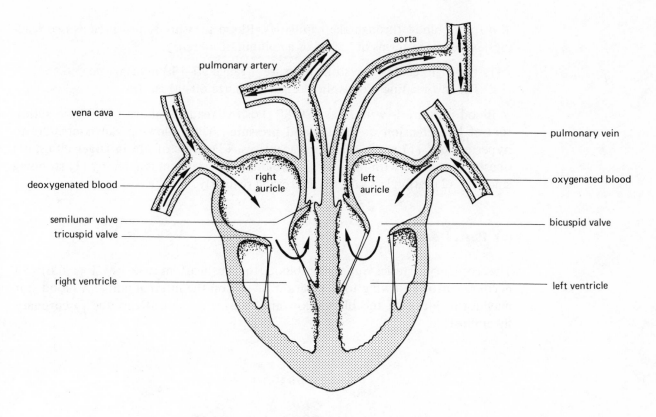

Fig. 6.5   The circulation of blood through the heart

6.3   **The Heart**

The heart beats 72–75 times per minute, when the body is resting, but can increase up to 200 per minute during violent exercise. The rate is faster during fevers and slower during sleep. 'Normal' resting heart rate covers a wide range from 50 to 110 beats per minute. 'Normal' rate means that it is the value found in the majority of people, i.e. mean or average.

The heart possesses inherent muscular activity. Although connected by nerves to the brain, it will continue to beat after these nerves are severed.

(a)   Sequence of Heart Beat

1. **Systole** – atria contract simultaneously; blood is expelled into the ventricles and then the ventricles contract simultaneously (i.e. **atrial** followed by **ventricular systole**).
2. (a) Blood in the ventricles is forced into the pulmonary arteries and the aorta. Semilunar valves at the base of each vessel prevent return of blood to the ventricles when they relax.
   (b) Blood is forced against the bicuspid and tricuspid valves which close, preventing backflow of blood into the atria.
3. **Diastole** is a pause during which all parts of the heart relax.

(b)   **Blood Pressure**

High blood pressure is maintained in the circulatory system by the pumping action of the heart and this forces the blood around at a high speed. This pressure also forces blood through the capillaries. Blood pressure is measured as two readings expressed in terms of height of a column of mercury:

1. The high reading is systolic pressure — range 90–140 mmHg.
2. The low reading is diastolic pressure — range 60–90 mmHg.

Blood pressure is written as 120/80. Positive readings above 140/90 are indicators of **hypertension** or '**high blood pressure**'. The following can contribute to hypertension: (1) some oral contraceptives; (2) stress of life in larger cities; (3) obesity associated with fatty diets. Avoiding hypertension is aided by (1) stopping smoking; (2) eating less salt; (3) exercising regularly.

(c) Heart Disease

The coronary arteries supplying blood to the heart muscle (see Fig. 6.6) can become narrowed owing to deposition of fats on the internal lining. A blood clot may form, blocking the blood flow and causing a heart attack due to **coronary thrombosis**.

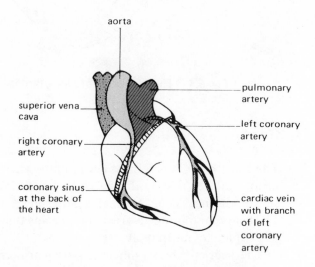

Fig. 6.6   Blood vessels of the heart

Arteriosclerosis is a general term covering degeneration of blood vessels. The commonest form is **atheroma** or **atherosclerosis**, which is due to the degeneration of the lining of the walls of an artery causing a blood clot to form. This obstruction means that less oxygen gets to the tissues supplied and thus they may be destroyed or damaged (see Table 6.3). The causes of deposits on artery walls is still not clear, but a connection has been established between **cholesterol** (a fatty substance) present in the blood and the incidence of the disease. Other important factors in the search for a cause of atherosclerosis are:

1. **Reduction in cholesterol level** by diet changes can alter the outcome of possible coronary disease.
2. **Heredity** seems to be important, for the death of one parent from coronary disease increases the chance of heart attacks by 100%.
3. **Heavy smokers** are much more likely to develop general atherosclerosis.
4. **Psychological and emotional stress** can be an important factor.
5. **Age** is a factor, since the chances of suffering coronary atherosclerosis increase the older we get.

Table 6.3 Causes of death

(a) Death from various diseases of the heart and blood vessels in 1954 and 1964, England and Wales			**(b) Ten major causes of death (% value of all causes) in 1965, England and Wales**	

Cause of death	*1954*	*1964*	*Cause of death*	*% value of all causes*
Rheumatic fever	299	61	1. Arteriosclerotic heart disease (including coronary disease)	21
Chronic rheumatic heart disease (the aftermath of an attack of rheumatic fever)	8 596	6 171	2. Cancer	19.5
Arteriosclerotic and degenerative disease of the heart			3. Vascular disease affecting the central nervous system	14
1. arteriosclerosis of coronary arteries	67 884	106 290	4. Diseases of the respiratory system	12
2. other causes of degenerative heart diseases	60 162	34 419	5. Accidents (including suicide)	4.5
Hypertensive disease (raised blood pressure)	20 573	13 195	6. Diseases of the digestive system	2.5
Disease of arteries	12 043	14 989	7. Diseases of the urogenital system	} 1
Generalised arteriosclerosis	10 035	10 911	8. Congenital malformations	
			9. Diabetes mellitus	} 0.5
			10. Tuberculosis	

6.4 The circulation

For a diagram of the circulation of blood, see Fig. 6.4.

The circulation is said to be double, consisting of (1) the **pulmonary circulation**, in which the blood travels from heart to lungs and back to the heart again, and (2) the **body (systemic) circulation**, in which the blood travels from the heart to all the organs of the body (except the lungs) and back to the heart again.

For the differences between arteries and veins, see Table 6.4.

Table 6.4 Differences between arteries and veins

Arteries	*Veins*
1. Carry blood away from the heart	Carry blood towards the heart
2. Carry blood with a high oxygen content (except the pulmonary artery)	Carry blood with a low oxygen content (except the pulmonary vein)
3. Walls are thick, muscular and elastic	Walls are thin (little muscle)
4. Valves are absent (except at the base of large arteries leaving the heart)	Valves are present throughout their length
5. Blood flows rapidly under high pressure	Blood flows slowly under low pressure
6. Blood flows in pulses	Blood flows smoothly
7. Tend to lie deeper in the body	Tend to lie near the body surface

Capillaries are minute vessels with walls one cell thick and only wide enough to permit one red blood cell to pass at a time (see Fig. 6.2). Water and dissolved substances pass in and out of the capillary walls. White blood cells can pass between the cells through the capillary walls.

Blood returns from the tissues to the heart through the veins at low pressure after passing through the capillaries. The blood is pushed through the veins by the contraction of surrounding skeletal muscle. Backflow is prevented by the semilunar valves in the veins. The venous blood emerges from the major organs with less oxygen, more carbon dioxide and more nitrogenous waste. The exception is the pulmonary veins, returning oxygenated blood from the lungs to the heart.

Varicose veins are a result of Man's upright stance, and develop through weakness at the semilunar valves in the major veins of the legs.

(a) Foetal Circulation

The circulation of blood in the foetus differs slightly from that in the adult. There is no functioning of the lungs in the foetus, and therefore the pulmonary circulation is not important until after birth. Two structural changes short-circuit the lungs: (1) a hole in the partition between left and right atria; (2) a small connecting channel between the aorta and the pulmonary artery. At birth both of these bypass mechanisms close up, so directing the blood through the lungs.

6.5 The lymphatic system

For a diagram of the lymphatic system and lymph glands (nodes), see Fig. 6.7.

The function of this system is to return tissue fluid from the tissues to the main circulatory system. The tissue fluid is formed mainly through plasma escaping from the capillaries. The tissue fluid on entering the lymphatic system becomes lymph. Small lymph vessels in the tissues join together and form the two main lymphatic ducts. These return the lymph to the main blood system through an opening in the subclavian veins.

Flow of lymph is brought about by the action of muscles pressing on the vessels. Semilunar valves (as in veins) prevent backflow. Lymph nodes present in the neck, groin and armpit are important in defending the body against micro-organisms by (1) producing lymphocytes responsible for antibody production, and (2) acting as a filter for micro-organisms in the lymph.

6.6 Questions and answers

(a) Multiple-choice Questions

1 When a blood smear is prepared, a drop of blood is drawn over the surface of the microscope slide by the leading edge of another slide. Which one of the following is the reason for this technique?

A It prevents the red blood cells from being crushed by the edge of the slide.

B It ensures that the red and white blood cells are stuck to the slide.

C It produces a film of blood one cell thick.

D It prevents the white blood cells from being crushed by the edge of the slide.

E It ensures that the blood will dry out quickly for staining.

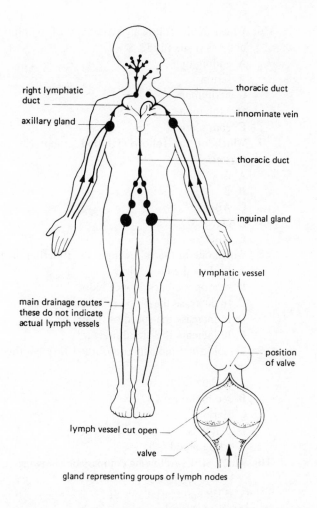

right lymphatic duct

axillary gland

thoracic duct

innominate vein

thoracic duct

inguinal gland

lymphatic vessel

main drainage routes — these do not indicate actual lymph vessels

position of valve

lymph vessel cut open

valve

gland representing groups of lymph nodes

Fig. 6.7   Position of the main lymph glands and vessels

2   Which of the following blood groups can be transfused into any one person?

   A A

   B B

   C AB

   D O

   E ABO

3   Which of the following are essential for the clotting of human blood?

   A potassium ions

   B calcium ions

   C sodium ions

   D iodine ions

   E fluorine ions

4   Which one of the following is not the function of lymph?

   A digesting fats

   B transporting hormones

   C carrying white blood cells

   D collecting urea

   E carrying chloride

5   The proportion of red blood cells to white blood cells in human blood is

   A 6000 : 1

   B 500 : 1

   C 5 000 000 : 1

   D 10 000 : 1

   E 10 : 1

6 Which of the following passes out of the blood capillaries and surrounds the cells of the body as tissue fluid?

A endolymph

B serum

C perilymph

D exolymph

E endoplasm

7 Which of the following blood groups can receive blood transfusion from any other person?

A A

B B

C AB

D O

E ABO

8 Which one of the following is the function of the valves in the heart?

A to slow the blood flow

B to stop the backflow of blood

C to decrease the blood pressure

D to increase the speed of flow

E to increase the blood pressure

9 Soluble food molecules absorbed through the wall of the small intestine are transported by

A plasma

B red blood cells

C fibrin

D platelets

E white blood cells

10 Which of the following comparisons between arteries and veins is **not** correct?

	Arteries	*Veins*
A	thick and muscular wall	thin wall
B	blood flows in pulses	blood flows smoothly
C	valves present	valves absent
D	blood with low carbon dioxide levels	blood with high carbon dioxide levels
E	blood at high pressure	blood at low pressure

11 Which one of the following arteries carries deoxygenated blood?

A renal artery

B hepatic artery

C carotid artery

D coronary artery

E pulmonary artery

12 The blood vessels to the mammalian foetus are not connected directly to those of the mother. Which of the following helps to explain this?

A Foetal blood is entirely different in composition from that of the mother.

B If the systems were connected, the direction of flow in the foetal system would be reversed.

C Food in the bloodstream of the mother is unsuitable for the foetus.

D Blood pressure in the mother's blood vessels is much higher than that of the foetal blood vessels.

E The foetus is not using its blood vessels, since it has no need to breathe in the womb.

13 Which of the following is unique in healthy heart muscle?

A only contracts

B well supplied with blood

C both expands and contracts

D never contracts, only relaxes

E never tires

14 Which of the following is characteristic only of arteries?

A Valves are present.

B Contain red blood cells.

C Carry blood away from the heart.

D Contain deoxygenated blood.

E Contain white blood cells.

15 Which of the following is the function of the semilunar valves in pulmonary artery and aorta?

A to separate atria (auricles) from ventricles

B to maintain the pulse in the arteries

C to close off the exits from the heart when it contracts

D to prevent backflow of blood into the ventricles

E to increase arterial pressure

16 Which of the following is produced in the lymph nodes?

A fibrinogen

B red blood cells

C blood platelets

D some white blood cells

E red blood cells and white blood cells

17 Which of the following is the function of the blood platelets?

A transport of hormones

B maintenance of blood flow

C carrying carbon dioxide

D helping blood to clot

E maintenance of blood group type

(b) Structured Questions

1 The table shows the number of red blood cells in the blood of people regularly living at different altitudes.

Altitude above sea level (metres)	Number of red blood cells (per mm³ of blood)
0 (sea level)	5 000 000
1000	6 000 000
3000	7 000 000
5500	7 500 000

(a) What is the function of red blood cells? (one line) (1)

(b) What is the effect on the blood of a person living at 1000 metres or more above sea level? (one line) (1)

(c) Suggest why this change in the blood helps a person live at a high altitude (two lines) (2)

(SREB)

2 (a) The graph below shows the changes recorded in heart beats per minute under a variety of different circumstances.

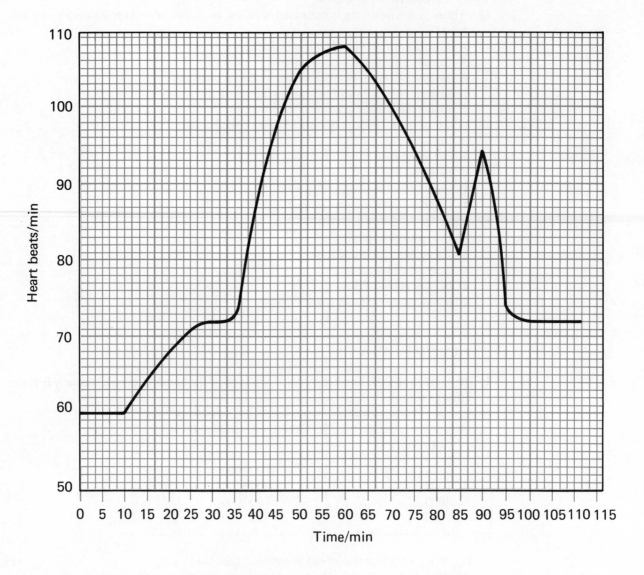

(i) From the graph above give the period of time when the man was:

    1. frightened by a loud bang, ...

    2. sleeping, ...

    3. running, ...

    4. waking up.   ....

                                                           (4)

(ii) What is the man's normal pulse rate when awake but at rest? (one line)   **(1)**

(iii) Which hormone is responsible for increasing the heart rate? (one line)   **(1)**

(b) The diagram below shows a front view of the heart cut lengthways.

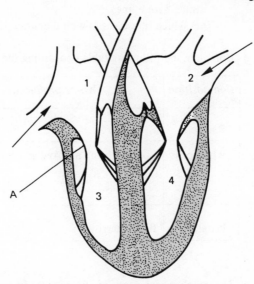

(i) What is the function of the part labelled **A**? (one line) **(1)**

(ii) In which two chambers is blood containing most oxygen found? (one line) **(1)**

(iii) Which chamber produces the greatest pressure? (one line) **(1)**

(iv) What is the name of the muscle in the wall of the heart? What special property does it possess?

Name (one line)

Special property (two lines) **(2)**

(c) (i) When a person takes an aspirin for a headache, the drug is carried by the blood from the gut to the head. Using the names on the diagram below, list, in the **correct** order, the blood vessels the drug passes through from when it leaves the gut until it reaches the head, by the shortest route.

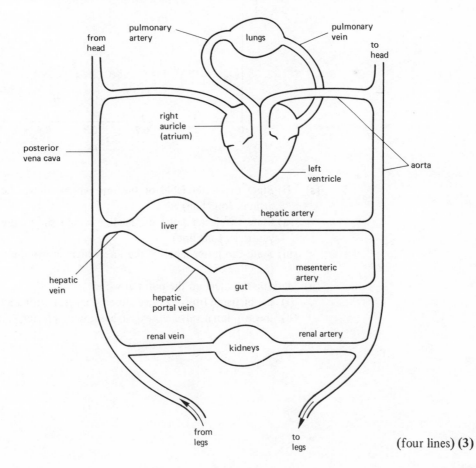

(four lines) **(3)**

87

(ii) Which blood vessel on the diagram contains the highest level of sugar after a meal? (one line) **(1)**

(iii) Which blood vessel would contain the highest level of oxygen? (one line) **(1)**

(YHREB, 1985)

3 Complete the following table concerning the composition and function of blood.

Composition	Where produced	One function
Red cells	. . . .	. . . .
White cells	Some in bone marrow and some in	. . . .
Platelets	Bone marrow	. . . .

(5)

(EAREB)

4 The graph shows some of the changes in pressure that occur within the left atrium, the left ventricle and the aorta during a single heart cycle.

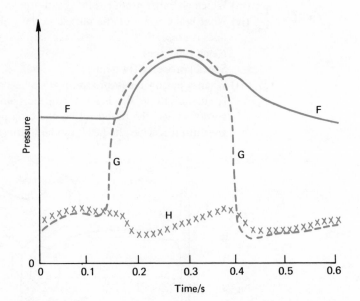

(a) (i) State the letter label of the line which shows the pressure changes in the left atrium. (one line)

(ii) State the letter label of the line which shows the pressure changes in the left ventricle. (one line)

(iii) State the letter label of the line which shows the pressure changes in the aorta. (one line) **(3)**

(b) Mark on the diagram the point at which

(i) the bicuspid (mitral) valve closes — label it with a letter **Y**.

(ii) the semilunar valve closes — label it with a letter **Z**. **(2)**

(AEB, 1984)

5

Blood of individual	X	Y	Z
Serum from blood of group A	clumps	clumps	no clumping
Serum from blood of group B	no clumping	clumps	no clumping

The blood of three individuals, **X**, **Y** and **Z**, was tested to determine their blood group in the **ABO** system. The tests, and their results, are illustrated in the figure above.

(a) State the blood group to which each individual belongs
   (i) **X** belongs to group . . . . . . *B* . . . .
   (ii) **Y** belongs to group . . . . *AB* . . . .
   (iii) **Z** belongs to group . . . . *O* . . . .

(b) What causes the red blood cells of individual **X** to change when the blood of **X** is

mixed with the serum from group A? . . . . *Antibody b present in the* . . . .
serum of group A

(c) To which of the blood groups in the **ABO** system must a donor belong to give a blood transfusion to individual **Z**?

Group O

(d) A donor has blood of group B. To which blood groups can his blood be given safely?

Groups B and AB

(e) Why is a person of blood group AB called a *universal recipient*?

No antibodies in serum. Therefore receive A, B, AB and O

(f) What is meant by the *compatibility* of blood when used in a blood transfusion?

That the donor blood can be received into the
blood of the recipient with no agglutination

(g) (i) What part do *platelets* in a person's blood play in the formation of a blood clot?

Release thrombokinase which acts on prothrombin
to form thrombin

   (ii) What is *fibrin*? *Thrombin acts on fibrinogen to form*
fibres (fibrin) - (form a clot)

   (iii) Name the mineral ion that must be present in blood for a blood clot to be formed.

Calcium

(h) How do the processes of *clumping* (agglutination) and *clotting* differ when they take place in a specimen of blood?

   (i) In clumping . . *Due to red blood cells forming clumps*
when incompatible blood is mixed

89

(ii) In clotting *Due to formation of fibrin, which entangles red and white blood cells*

(UCLES)

6 The figure shows three types of cell found in human blood.

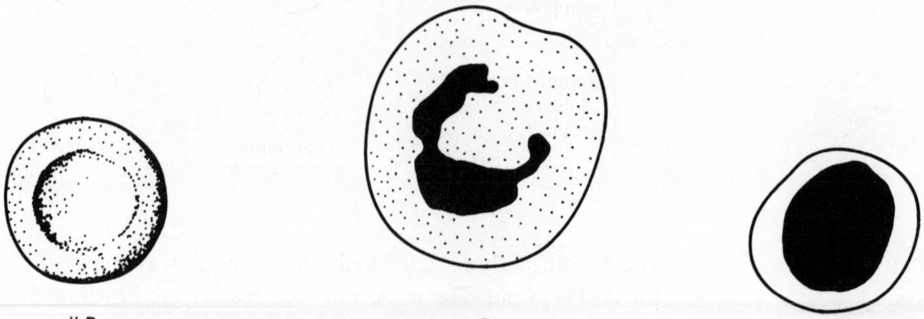

cell P   cell Q   cell R

(a) (i) What type of blood cell is **Q**?

Leucocyte (white blood cell) polymorph **(1)**

(ii) State **two** features of cell **Q** which enabled you to distinguish it from the other cells.

1 *Lobed nucleus*

2 *Granular cytoplasm* **(1)**

(iii) Explain the function of this type of blood cell.

It provides protection against the invasion of the body by bacteria. These are engulfed by the cytoplasm (phagocytosis) **(1)**

(b) (i) Which **one** of the cells shown in the diagram would contain haemoglobin?

P **(1)**

(ii) State the function of haemoglobin.

To chemically combine with oxygen (forms oxyhaemoglobin) **(1)**

(c) (i) Name the liquid part of human blood.

Plasma **(1)**

(ii) What is its colour when separated from blood cells?

Yellow **(1)**

(d) (i) Complete the table to show where the **named** substances enter and leave the blood.

Substance	Enters blood	Leaves blood
glucose	*Villi*	*Liver*
carbon dioxide	*Muscles*	*Alveoli*

(2)

(ii) Name **two** other substances transported by the liquid part of the blood.

1 *Hormones* . . . .

2 *Antibodies* . . . .

(1)

(UCLES)

7 (a) The figure is a histogram showing the rate of blood flow to various organs.

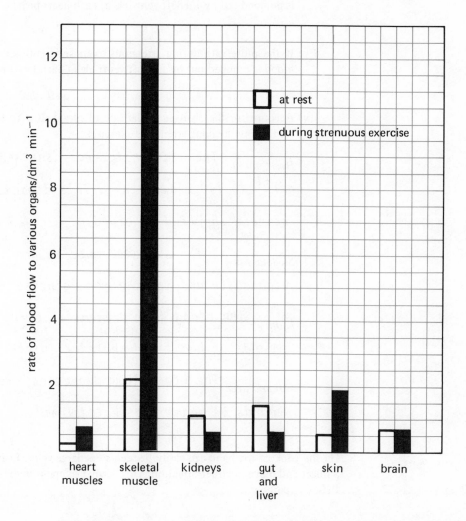

(i) Record, in the table below, the volume of blood flowing per minute to each of the organs when the body is at rest, and when the body is undergoing strenuous exercise.

	Volume of blood/dm³ min⁻¹	
	at rest	*during strenuous exercise*
Heart muscle	*0.2*	*0.7*
Skeletal muscle	*2.2*	*12.0*
Kidneys	*1.1*	*0.6*
Gut and liver	*1.4*	*0.6*
Skin	*0.5*	*1.9*
Brain	*0.7*	*0.7*

91

(ii) What is the total volume of blood per minute being pumped by the left ventricle to all of these organs when the body is:

at rest . . . . 6.1 dm^3 min^{-1} . . . .

undergoing strenuous exercise? . . 16.5 dm^3 min^{-1} . . . .

(iii) If the pulse rate when the body is at rest is 70 per minute, what volume of blood is pumped out by the left ventricle at each heart beat?

427 dm^3

(iv) If the pulse rate during strenuous exercise is 160 per minute, what volume of blood is pumped out by the left ventricle at each heart beat?

2640 dm^3

(v) Account for the changes in the rate of blood flow to the following organs as a result of undergoing strenuous exercise.

Heart muscle *The heart rate increases from about 80 beats* *min^{-1} to 140 beats min^{-1}. This demands more oxygen and* *glucose supplied by blood to heart muscle.*

Skeletal muscle *The skeletal muscles are contracting* *vigorously and require more glucose and oxygen.*

Kidney *The blood flow decreases, owing to the* *demands of heart and skeletal muscle.*

Gut and liver *The blood flow decreases, owing to the* *demands of heart and skeletal muscle.*

(UCLES)

8 On the back of my hand can easily be seen some large veins. I can **lightly press** them with a finger and move in either direction. This is what happens in each case.
Before experiment.

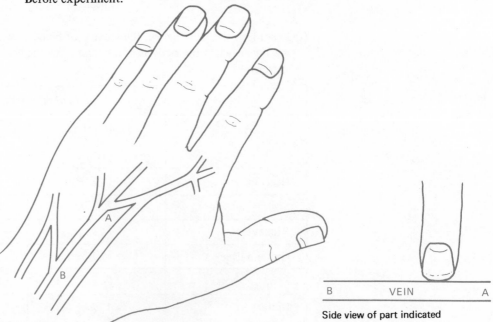

Back of **left** hand (note points A and B).

Side view of part indicated

B   VEIN   A

1. Moving finger from A to B and stopping at B.

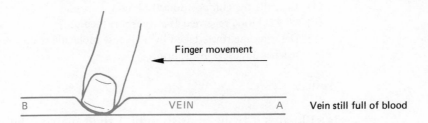

2. Moving finger from B to A and stopping at A.

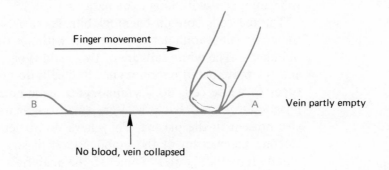

Explain what happened in each case.

1. From A to B the blood flowing towards the heart follows behind the finger and the vein remains full of blood.

2. From B to A the blood is pushed back but no blood follows the finger, since it is prevented from backflow by a semilunar valve at B.

(YHREB, 1984)

(a) Describe the composition of blood. **(6)**

(b) Why is blood regarded as a transport medium? **(8)**

(c) Describe the route taken by a glucose molecule from the time it enters the blood until it reaches a muscle in the arm. **(6)**

Answer

(a) Blood is a tissue consisting of a fluid matrix, the plasma, in which are found different dissolved substances. The diagram represents the different types of mammalian blood cell. [It would be worth while to insert here a diagram of blood cells as shown in Fig. 6.1. It is a simple and quick drawing which saves time compared with long explanations.]

The red cells contain haemoglobin. There are approximately five million of them per cubic millimetre, compared with six thousand white cells per cubic millimetre. The white cells are of two main types, the polymorphic phagocytes and the more ovoid lymphocytes. Red cells do not contain a nucleus, while all types of white cells do. Lymphocytes make antitoxins and antibodies, while phagocytes engulf bacteria. Very small fragments of cells called platelets are also present in the plasma. They have no nucleus and are concerned with the clotting mechanism of the blood. Blood plasma makes up about 55% of the blood. It is 95% water. Some of the materials contained in it are dissolved gases, proteins, hormones, enzymes and antibodies.

(b) Blood is regarded as a transport medium because it is able to convey materials from one part of the body to another by means of its circulation through blood vessels and capillaries. Red blood cells transport oxygen as oxyhaemo-globin from the lungs to metabolising tissues. They also carry some carbon dioxide back to the lungs. However, most of the carbon dioxide is transported as hydrogen carbonate (bicarbonate) in the plasma. The plasma also carries waste matter from the tissues to the excretory organs. Hormones produced by the ductless glands are circulated around the body in the plasma until they reach their respective target organs. In a similar manner the digested food from the ileum is conveyed first to the liver by the blood and then eventually around the body. Heat from the liver and muscles is also passed to all parts of the body. White blood cells and platelets, both used in the defence of the body, can also be conveyed to wounds, burns and other places where they can help protect or restore the tissues. Fibrinogen and prothrombin, proteins concerned in blood clotting, are carried by plasma to wounds.

(c) The route taken by a glucose molecule is as follows: (1) capillary in villus; (2) hepatic portal vein; (3) liver; (4) hepatic vein; (5) vena cava; (6) right auricle of heart to right ventricle; (7) pulmonary artery; (8) lungs; (9) pulmonary vein; (10) left auricle of heart to left ventricle; (11) aorta; (12) subclavian artery of arm.

Notes

1. Section (a) includes diagrams of blood cells which are adequate for providing information on the structure of these cells.

2. In section (b) a full description is given of the transport functions of blood. Note the mention of heat transfer, often forgotten by students.

3. Section (c) requires a clear recall of the parts of the circulatory system and the organs involved. The list should be as full as possible, since the examiner will stop marking at the first wrong vessel or organ.

(c) Answers to Objective and Structured Questions

(i) *Multiple-choice Questions*

1. C   2. D   3. B   4. A   5. B   6. B   7. C   8. B   9. A   10. C   11. E   12. D
13. E   14. C   15. D   16. D   17. D

(ii) *Structured Questions*

1 (a) To carry oxygen as oxyhaemoglobin.
  (b) There is an increase in the number of red blood cells per mm^3.
  (c) This enables uptake of sufficient oxygen, even though there is less oxygen
      in the rarefied atmosphere of high altitudes.

2 (a) (i)   1 at 85 to 96 minutes
             2 at 0 to 10 minutes
             3 at 35 to 60 minutes
             4 at 10 to 30 minutes
      (ii) 72 beats per minute
      (iii) Adrenalin
  (b) (i) To prevent backflow of blood on contraction of the ventricle (chamber
          3)
      (ii) 2 and 4
      (iii) 4
      (iv) Cardiac muscle: continuous beat (not subject to fatigue)
           or: Has an intrinsic stimulation of beat
  (c) (i) Hepatic portal vein; hepatic vein; posterior vena cava; (heart) pulmon-
          ary artery; (lungs) pulmonary vein; (heart) aorta
      (ii) Hepatic portal vein   (iii) Pulmonary vein

3

Where produced	One function
in bone marrow	carry oxygen as oxyhaemoglobin
in lymphoid tissue	engulf bacteria
(bone marrow)	initiate clotting of blood

4 (a) (i) H   (ii) G   (iii) F
  (b) (i)   At first junction of line G with line H–Y
      (ii)   At second junction of line G with line F–Z

Questions **5–8** have the answers supplied with the questions.

7 Breathing Mechanism and Respiration

7.1 Breathing and respiration

Breathing is the act of gaseous exchange involving breathing out (exhaling) and breathing in (inhaling) using the rib cage, the intercostal muscles and the diaphragm muscles. It is sometimes called external respiration but this term should **not** be used.

Respiration is the chemical process involving the oxidation of food substances in the cells to release energy (also termed cell, tissue and internal respiration).

The gaseous exchange system requires:

1. A **medium** − in which the required gases are present.
2. A **respiratory surface** − a large moist surface area where the gases are dissolved before diffusing in and out of the cells.
3. A **transport system** − a system to move the dissolved gases to all cells and to collect gases needing to be eliminated.

For the structure of the lungs and associated organs, see Figs. 7.1 and 7.2.

Man inhales through the nose (the mouth and buccal cavity should not be used). Air is **warmed** (by surface capillaries) and **filtered** (by mucus and ciliated epithelium) when passing through the narrow air-channels of the nasal passage. Sense cells sensitive to chemicals (see Chapter 10) lie in the upper surface of the nasal cavity.

The **trachea** and the **bronchi** are also lined by ciliated epithelium and mucus-secreting cells. The mucus collects dust and bacteria, while the beating cilia move the mucus up to the throat, where it is swallowed.

The **larynx** joins the back of the throat (**pharynx**) by an opening called the **glottis**, which can be closed by a flap, the **epiglottis**. This prevents food entering the larynx and the trachea (see page 62). The larynx contains two folds of tissue, the **vocal chords**. These vibrate when air is expelled over them and, together with the resonance of the buccal cavity, make a variety of sounds.

The nasal cavity is separated from the buccal cavity by the **hard and soft palates**. Thus, when food is filling the buccal cavity and is being chewed, we can still breathe.

The trachea and the bronchi are strengthened by incomplete hoops of **cartilage** which prevent them collapsing when the neck bends. From the bronchi, a series of finely divided tubes (**bronchioles**) reach the tiny air-sacs or **alveoli**. These give an enormous surface area for gaseous exchange. They have a wall of one-cell thickness and are well supplied with blood capillaries (see Figs. 7.1 and 7.3). In the human lung there are approximately 300 million alveoli, providing an area about the size of a tennis court.

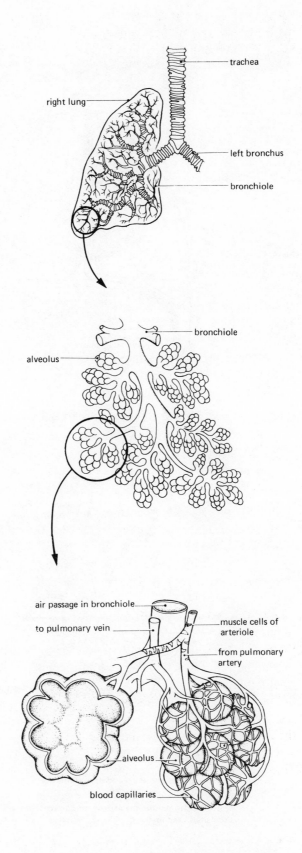

Fig. 7.1 The lung structure at three stages of magnification (*after D. G. Mackean*)

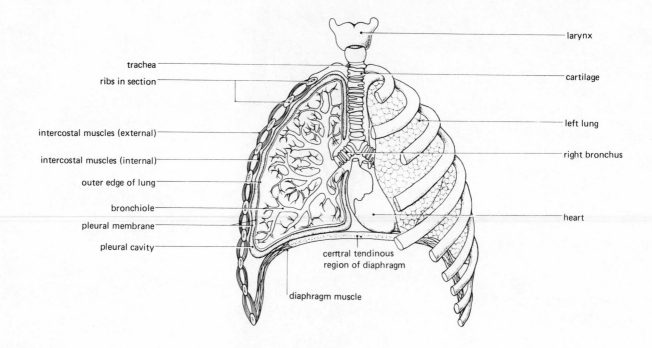

Fig. 7.2 The lungs, heart and associated organs

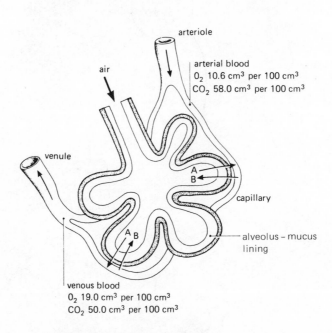

At A, O_2 diffuses into the blood through mucus, alveolar wall and capillary wall

At B, CO_2 diffuses into the alveolus through capillary wall, alveolar wall and mucus

Fig. 7.3 The exchange of gases in the alveoli

7.2　The mechanism of breathing

See Table 7.1 and Fig. 7.4.

Table 7.1　Comparison of mechanism of inhalation and exhalation

Inhalation	Exhalation
1. External intercostal muscles contract	External intercostal muscles relax
2. Internal intercostal muscles relax	Internal intercostal muscles contract
3. Ribs raised in Man	Ribs lowered in Man
4. Diaphragm contracts	Diaphragm relaxes
5. Diaphragm flattens	Diaphragm arches upwards
6. (by 3 and 5) Volume of thorax increases	(by 3 and 5) Volume of thorax decreases
7. Air pressure decreases	Air pressure increases
8. Air moves into the lungs	Air forced out of the lungs

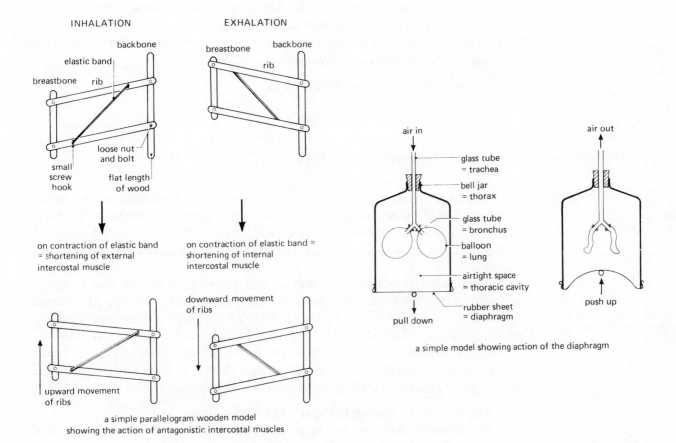

Fig. 7.4　Models to demonstrate breathing mechanisms

The external and internal intercostal muscles act as antagonistic pairs (see Chapter 9). Their action can be demonstrated by a simple model (see Fig. 7.4). Figure 7.4 also shows the bell-jar model of the chest, demonstrating the way in which the movement of the diaphragm alters the volume of the chest cavity. This change in volume changes the pressure, and air either moves in (diaphragm flattens) or moves out (diaphragm rises upwards). Like all biological models, there are a number of disadvantages or defects. Note that:

1. **No rib movement** is shown by the rigid glass walls.
2. The **balloons are too small** compared with the lungs which fill the chest cavity.
3. From 2 above, it follows that the **space** between balloons and wall **is too large**.
4. The action of the rubber sheet in relation to the diaphragm is incorrect. The diaphragm flattens on inhalation but is never pulled down below the horizontal, as in the model.

(a) Other Movements of Thorax and Diaphragm

1. **Hiccough** — a spasmodic contraction of the diaphragm forcing air out of the lungs.
2. **Cough** — a forced exhalation to clear the trachea and the larynx of mucus containing dust and bacteria.
3. **Yawn** — a long inhalation of air due to a variety of causes which may have resulted in carbon dioxide increase in the blood.
4. **Sigh** — a longer inhalation and exhalation than is usual in breathing, involving the thorax and the diaphragm.

For gas exchange in the alveoli, see Fig. 7.3.

7.3   Hygiene of the breathing mechanism

The violent expulsion of air during a **sneeze** carries thousands of droplets of moisture. The droplets are projected several metres and remain suspended, finally drying as dust. They contain bacteria and viruses, and these can be inhaled by another person, thus spreading infection (influenza, polio, pneumonia, whooping cough and scarlet fever are spread in this way).

Air pollution from industrial sources and the engines of cars and lorries is a problem of industrial countries. Soot, sulphur dioxide, carbon monoxide, nitrogen oxides and hydrocarbons are all present in the air we breathe. Respiratory ailments are both caused and aggravated by these substances (see Chapters 14 and 16).

Tobacco smoke is another major pollutant, which contains benzopyrene, a cancer-producing agent. It has been shown that there is a clear **connection between smoking and lung cancer**. Many investigations have demonstrated that:

1. Smokers have a **shorter life span** than non-smokers.
2. Incidence of cancer of the lungs is associated with **the number of cigarettes** smoked per day.
3. **Chronic bronchitis and emphysema** are both brought on by excessive smoking.
4. There is a higher incidence of **disease of heart and blood vessels** among smokers.
5. Smoking may affect health in other ways, e.g. produce **cancer** of the mouth, the pharynx and the oesophagus.
6. Smoking during pregnancy **reduces the birth weight** of the baby.

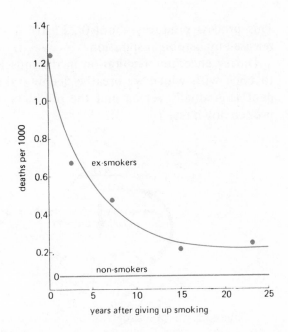

Fig. 7.5 Death rate from lung cancer after giving up smoking, compared with death rate from lung cancer in non-smokers

Asthma – more than half of the cases of this common disease is caused by sensitivity to atmospheric pollen, dust, fur, etc. The difficulties of breathing are caused by contraction of muscle cells resulting in constriction of airways.

7.4 Respiration

By using radioactive tracers in pure foods fed to laboratory animals it has been shown that carbon dioxide (CO_2) produced by breathing has come from food. Radioactive carbon (^{14}C) in glucose, $^{14}C_6H_{12}O_6$, results in the production of $^{14}CO_2$. The sugar has been broken down by **aerobic respiration**:

$$^{14}C_6H_{12}O_6 + 6O_2 \rightarrow 6H_2O + 6^{14}CO_2 + energy$$

glucose oxygen water carbon
                                      dioxide

The importance of this breakdown is the release of energy, vital to all living processes. Each small quantity of energy released is stored in the chemical compound **ATP** (adenosine triphosphate), which is formed from the related molecule **ADP** (adenosine diphosphate) and phosphate:

$$ADP + phosphate + energy \rightarrow ATP$$

When energy is released from ATP, ADP and phosphate are formed. For the oxidation of each molecule of glucose, 36 ATP molecules are formed with a total energy value of 2.81×10^6 J = 2810 kJ. The 36 ATP molecules represent about 39% of total available energy in the glucose molecule.

At times of vigorous activity (e.g. running rapidly), energy can be produced in the absence of oxygen, i.e. **anaerobic respiration**. The activity of the muscles outstrips the supply of oxygen available. Glucose is then partly broken down to lactic acid, with the release of a limited amount of energy:

$$C_6H_{12}O_6 \rightarrow 2(C_3H_6O_3) + energy$$

glucose lactic acid

101

This process produces about 0.22×10^6 J = 220 kJ, i.e. about 8% of the energy released by aerobic respiration.

During anaerobic respiration in the muscles, there is built up an **oxygen debt**, to cope with which we breathe deeply, taking in great gulps of air. The oxygen debt is gradually repaid and the products of anaerobic respiration are gradually broken down (see Fig. 7.6).

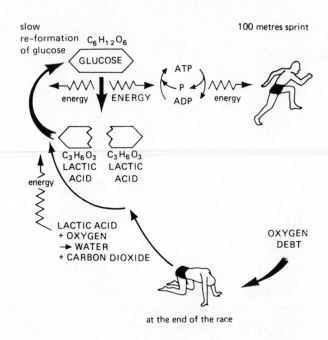

Fig. 7.6   Release of energy in the muscles of a man running a 100 metres sprint

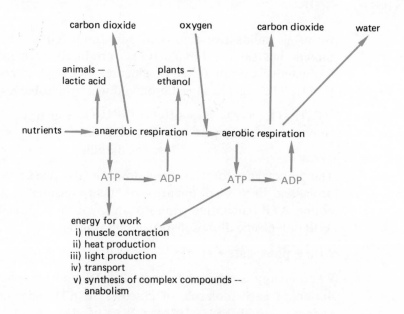

Fig. 7.7   Summary of energy flow in respiration

7.5   Volume of air moved into and out of the lungs

The body under normal resting conditions moves only a small volume of air into and out of the lungs. This is called the **resting tidal volume** and is about 500 cm³.
A **forced inhalation**, to its fullest extent, can take in an additional 2000 cm³. This

is the **inspiratory reserve volume** and together with the tidal volume forms the **inspiratory capacity**. The **expiratory reserve volume** can be breathed out after a normal gentle exhalation, and is about 1300 cm³. The total amount of air that can be forced out after the deepest possible inhalation is the **vital capacity**. After this there is still a **residual volume** left in the lungs, amounting to about 1500 cm³ (see Fig. 7.8).

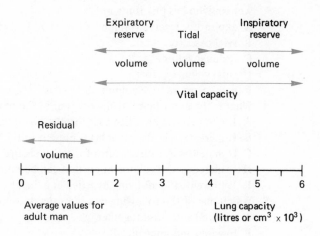

Fig. 7.8

7.6   Content of inhaled and exhaled air

See Table 7.2, which gives the data for the resting condition. After exercise, the oxygen consumption rises slightly and so does the carbon dioxide output. Note that the lungs are not very efficient in extracting oxygen, since exhaled air still has about 17% oxygen. This is why exhaled air can be used in mouth-to-mouth resuscitation.

Table 7.2   Gaseous content of inhaled and exhaled air

Gas	Inhaled air (%)	Exhaled air (%)
Carbon dioxide	0.03	3.50
Oxygen	20.93	16.89
Nitrogen	79.04	79.61
	100.00	100.00

7.7   Questions and answers

(a) Multiple-choice Questions

1 Which of the following is **not** true for the breathing process in Man?

	Inhalation	Exhalation
A	ribs raised	ribs lowered
B	diaphragm flattens	diaphragm arches upwards
C	volume of thorax decreases	volume of thorax increases
D	air pressure decreases	air pressure increases
E	air rushes in	air is forced out

2 Which one of the following describes the tidal volume of air in human lungs?
  A a forced inhalation that takes in about 2000 cm^3 of air
  B the volume of air moved in and out of the lungs at rest
  C the total volume of air that can be contained in the lungs
  D a gentle exhalation that forces out about 1300 cm^3 of air
  E the volume of air that is present in the lungs after exhalation

3 Which one of the following represents the air moved into and out of the lungs of a woman who has just run a mile?
  A expiratory reserve volume
  B vital capacity
  C residual volume
  D tidal volume
  E inspiratory reserve volume

4 What is the importance of the division of the lungs into alveoli?
  A It allows more blood to pass through the lung capillaries.
  B It increases the time that air is present in the lungs.
  C It increases the surface area for the exchange of gases.
  D It increases the volume of air that can be taken in.
  E It prevents the air being expelled at exhalation.

5 Which one of the following describes the movement of ribs during exhalation in Man?
  A lowered and close together
  B lowered and unchanged
  C raised and farther apart
  D raised and close together
  E lowered and farther apart

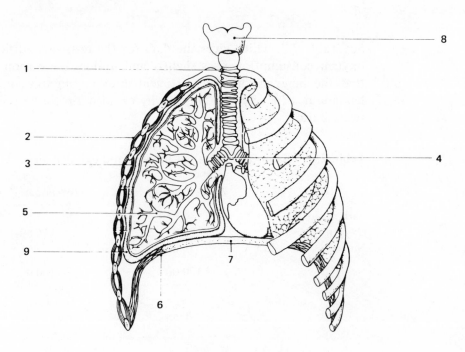

Fig. 7.9

Examine Fig. 7.9 and answer questions **6–11**.

6 Which of the following contract on inspiration?
  A 2 and 7
  B 3 and 6
  C 2 and 6
  D 3 and 7
  E 2 and 9

7 What is the name of structure 4?
   A trachea
   B alveolus
   C bronchus
   D air sac
   E bronchiole

8 Which of the following contract(s) on expiration?
   A 2
   B 3 and 6
   C 3
   D 2 and 6
   E 3 and 7

9 Which one of the following is directly concerned with sound production?
   A 1
   B 4
   C 5
   D 8
   E 9

10 Tube 1 is lined with ciliated epithelium. Which of the following is the function of the cilia?
   A to prevent food particles passing to the lungs
   B to present a large surface area for the absorption of gases
   C to absorb oxygen and release carbon dioxide
   D to beat towards the pharynx, moving mucus and dust
   E to secrete mucus and thus moisten the incoming air

11 Which of the following events take place during inspiration (inhalation)?

	muscle 2	muscle at 6	internal pressure at 5
A	relaxes	contracts	increases
B	contracts	relaxes	increases
C	relaxes	relaxes	decreases
D	contracts	contracts	decreases
E	relaxes	contracts	increases

12 When the environmental temperature is at $12°C$, which one of the following represents the temperature of the air breathed out?
   A $12°C$
   B $98.4°C$
   C $37.0°C$
   D $30.0°C$
   E $24.0°C$

13 Which one of the following word equations represents the process of respiration?
   A glucose → carbon dioxide + water + energy
   B glucose + carbon dioxide → oxygen + water + energy
   C glucose + oxygen → carbon dioxide + water + energy
   D oxygen + water → carbon dioxide + energy
   E water + carbon dioxide → oxygen + water + energy

14 During anaerobic respiration in plant cells the breakdown of glucose produces
   A alcohol and water
   B water and carbon dioxide
   C alcohol and carbon dioxide
   D carbon dioxide and lactic acid
   E oxygen and alcohol

15 Which one of the following is responsible for the conversion of sugar to alcohol?
   A viruses
   B yeast
   C oxygen
   D acid
   E carbon dioxide

16 Which one of the following forms lactic acid in muscles?

    A running

    B laughing

    C sleeping

    D fasting

    E dieting

17 Smoking tobacco reduces the amount of oxygen carried by the blood because

    A smoke contains carbon monoxide, which combines with haemoglobin.

    B tobacco ash collects in the lungs, so they can absorb less oxygen from the air.

    C cancer of the blood cells is caused.

    D smoke contains carbon dioxide, which destroys red blood cells.

    E tars in the smoke collect on the inside of veins, so helping to block them.

(SREB)

(b) Structured Questions

1 (a) Respiration may be defined as 'the oxidation of simple carbohydrates within the cells of living organisms, in order to release the energy necessary for life'.

Express this definition in terms of a word equation OR chemical equation. (two lines) **(4)**

In what way does photosynthesis help to maintain a balance of gases in the atmosphere? (three lines) **(2)**

(b) Complete the following table to show what physical changes take place in the thorax in order that we may breathe in and out.

	Inhalation (in)	*Exhalation (out)*
Ribs		
Diaphragm		
Lungs		

(c) Construct a diagram of an apparatus that could be used to investigate the difference between inhaled and exhaled air in terms of carbon dioxide content. Write a short explanation underneath the diagram describing the procedure, expected results and conclusion. (15 cm of space)

Procedure (three lines)

Expected results (two lines)

Conclusion (two lines) **(8)**

(EAREB)

2 Study the charts below, which show the deaths from diseases related to smoking.

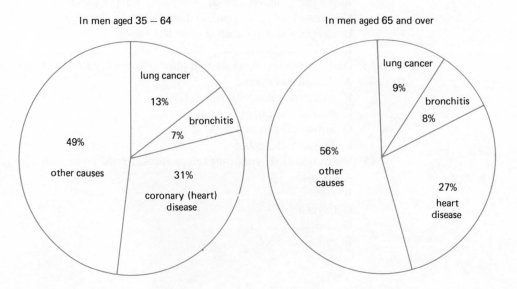

In men aged 35 – 64

lung cancer 13%

bronchitis 7%

49% other causes

31% coronary (heart) disease

In men aged 65 and over

lung cancer 9%

bronchitis 8%

56% other causes

27% heart disease

(a) What percentage of men aged 65 and over die from heart disease? (one line) **(1)**

(b) What percentage of men die as a result of other causes aged
  (i) between 35 and 64? (one line) **(1)**
  (ii) 65 and over? (one line) **(1)**

(c) What is the total percentage of deaths from lung cancer and bronchitis?
  (i) between 35 and 64? (one line) **(1)**
  (ii) 65 and over? (one line) **(1)**

(d) Give two other causes of heart disease.
  1 (one line) **(1)**
  2 (one line) **(1)**

(e) Give **one** reason why the deaths from other causes increase after the age of 64. (one line) **(1)**

(f) Give **two** biological reasons why it is a disadvantage for pregnant women to smoke.
  1 (one line) **(1)**
  2 (one line) **(1)**

(EMREB)

3 Figure (a) represents some of the organs found in the human thorax and Fig. (b) an enlargement of one of these organs.
  (a) Label the parts **A** to **J**. **(10)**

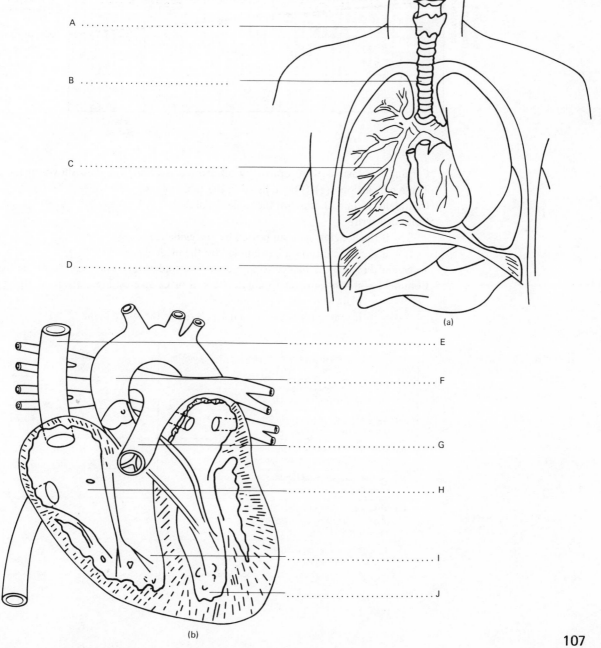

A ....

B ....

C ....

D ....

(a)

E ....

F ....

G ....

H ....

I ....

J ....

(b)

(b) (i) The ribs have not been shown in Fig. (a).
What is the name of the muscles attached to the ribs which assist in the breathing process? (one line) **(1)**
(ii) What part does the organ labelled **D** play in breathing? (four lines) **(4)**
(c) Refer to Fig. (b) and describe the circulation of blood from the right atrium (auricle) to the left atrium. (five lines) **(6)**
(d) Name two non-liquid constituents of blood. (two lines) **(2)**
(e) Name two dissolved constituents of blood. (two lines) **(2)**

(EAREB)

4 The figure below shows the breathing rate of a person at rest breathing different gas mixtures.

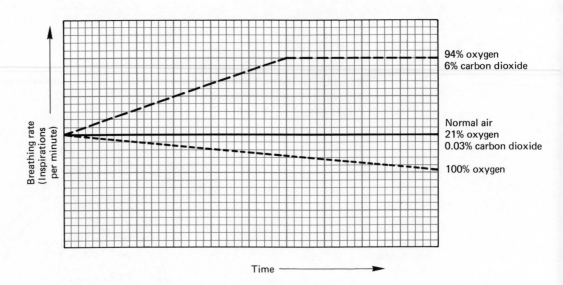

Certain conclusions were drawn by students concerning the data shown in the figure. These conclusions, (a)-(f), are listed in the table below.
Complete columns 1 and 2 in the table as follows:
Column 1 write
    S — if the conclusion is supported by the evidence
    C — if the conclusion is contradicted by the evidence
    N — if there is no evidence to support or contradict the conclusion from the figure
Column 2 — if your answer to column 1 is S or C, give evidence from the figure to explain your answer
        — if your answer to column 1 is N, leave the space blank

Conclusion	Column 1	Column 2
(a) The rate of breathing increases markedly in 6% carbon dioxide		
(b) A concentration of more than 6% carbon dioxide is fatal		
(c) There is no marked change of breathing rate under any conditions		
(d) Temperature is a major factor in increasing the breathing rate		
(e) The concentration of carbon dioxide plays an important part in governing the rate of breathing		
(f) Lack of oxygen is a stimulus to increased breathing rate		

(Total 12)

(L)

5 (a) The diagram represents an apparatus which can be used to demonstrate some of the features of the mechanism of breathing.

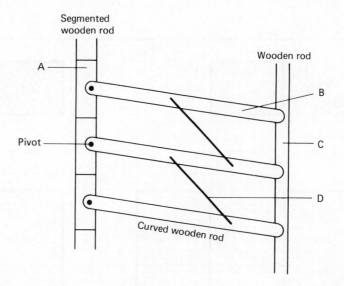

(i) Identify the structures of the thorax which are represented by the parts labelled **A**, **B** and **C**. **(3)**

(ii) Which muscle does part **D** represent? Explain concisely the effect of the contraction of this muscle during breathing. **(6)**

(b) Explain the functions of the thoracic skeleton, other than breathing. **(6)**

(c) The skeleton is built from *bone* and *cartilage*. Describe briefly **one** similarity and **one** difference between these two materials. **(3)**

(d) The table records some of the results of an investigation into breathing before and immediately following vigorous exercise.

	Before exercise	*After exercise*
Breathing rate (breaths per minute)	15	24
Volume of air expired per breath (dm^3)	0.5	1.5
Oxygen in inspired air (%)	21	21
Oxygen in expired air (%)	17	16

Showing the method of calculation,

(i) calculate the volume of air exchanged per minute before and after exercise, **(3)**

(ii) calculate the volume of oxygen absorbed per minute before and after exercise. **(3)**

(e) Explain why there is still an increased rate of oxygen absorption after exercise has finished. **(6)**

(AEB, 1985)

6 The bar graph (*histogram*) below shows the proportion of men aged 35 years who will die before they reach the age of 65. Study the graph and answer the questions about it.

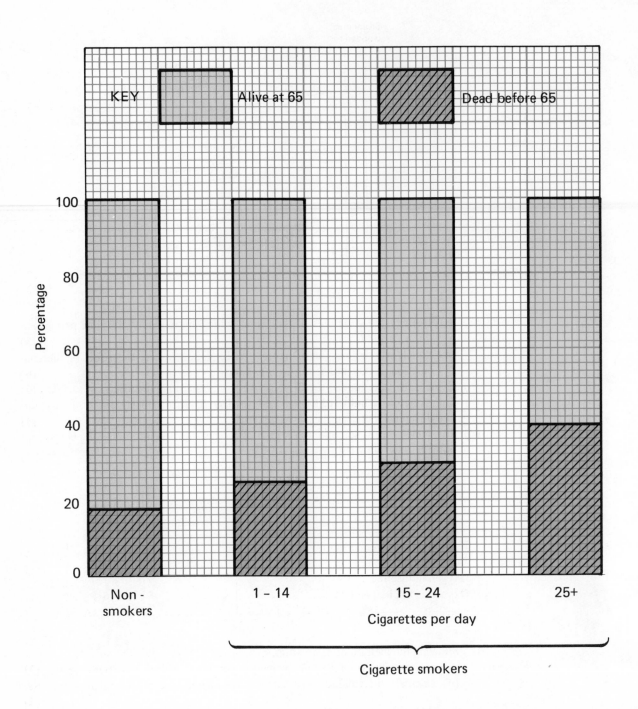

(a) What percentage of non-smokers die before the age of 65? (one line) **(1)**

(b) What percentage of smokers who smoke more than 25 cigarettes per day will die before the age of 65? (one line) **(1)**

(c) If a man aged 35 smokes 20 cigarettes per day, what is the percentage chance of surviving until the age of 65? (one line) **(1)**

(d) Name a substance (other than nicotine) produced by cigarettes, which harms the lungs. (one line) **(1)**

(e) Smokers are more likely than non-smokers to contract certain diseases of the lungs. Name two smoking-related diseases of the lungs. (two lines) **(2)**

(EMREB)

7 Study the charts below and answer the following questions.

Chart 1. Composition of air (%)

	Atmospheric air	Alveolar air	Exhaled air
Nitrogen	79.0	80.7	79.5
Oxygen	20.9	13.8	16.4
Carbon dioxide	0.03	5.5	4.1

Chart 2. Volume of each gas carried by 100 cm^3 of blood

	Blood entering lungs	Blood leaving lungs
Nitrogen	0.9 cm^3	0.9 cm^3
Oxygen	10.6 cm^3	19.0 cm^3
Carbon dioxide	58.0 cm^3	50.0 cm^3

(a) Which sample of air had the most carbon dioxide in it? (one line) **(1)**

(b) Which sample of air had the least carbon dioxide in it? (one line) **(1)**

(c) What is the percentage difference between the amount of carbon dioxide in alveolar air and in exhaled air? (one line) **(1)**

(d) What causes the difference in (c) above? (one line) **(1)**

(e) Which sample of air had most oxygen in it? (one line) **(1)**

(f) Which sample of air had least oxygen in it? (one line) **(1)**

(g) What is the percentage difference between the amount of oxygen in atmospheric air and alveolar air? (one line) **(1)**

(h) (i) Which blood sample contains the most carbon dioxide? (one line) **(1)**
    (ii) Which blood sample contains the least carbon dioxide? (one line) **(1)**

(j) What is the difference in volume of carbon dioxide (per 100 cm^3 of blood) between the blood samples? (one line) **(1)**

(k) Where has this extra carbon dioxide come from? (one line) **(1)**

(l) What is the difference in volume of oxygen (per 100 cm^3 of blood) between the blood samples? (one line) **(1)**

(m) Why is there a difference in the amount of oxygen between the blood samples? (one line) **(1)**

(n) Explain fully how air enters and leaves the lungs. (eight lines) **(8)**

(EMREB)

8 (a) In the mammalian thorax, when the diaphragm muscles relax, what change will take place to the air pressure inside the alveoli?

The air pressure will increase. **(1)**

(b) Explain briefly how the relaxation of the diaphragm muscles brings about this pressure change.

The diaphragm moves upwards, decreasing the volume of the thoracic cavity, thus increasing the pressure. **(1)**

111

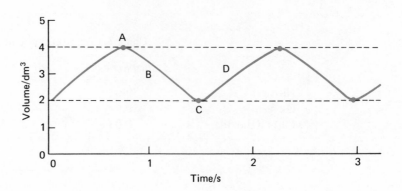

(a) The diagram in the figure shows the change in the volume of air in the lungs of a man during breathing. Use the diagram to answer the next four questions.
   (i) Which part of the graph **A, B, C** or **D** represents exhalation? .. *B*. ....
   (ii) How many breaths per minute is the man taking? . *40*. ....
   (iii) What is the change in the volume of air in the lungs during inspiration? . *2* . dm^3

   (iv) How do you know that the man is not resting? . *The amount of air* ....
      exchanged is 2 dm^3 = 2000 cm^3, i.e. too much for
      tidal air.

(b) What is *tidal air*? *The air breathed in and out at rest*
      (breathing quietly).

(c) What is *supplemental air*? *Also called expiratory reserve volume.*
      The amount of air in a forced expiration after tidal
      expiration.

(d) What is meant by *vital capacity* of lungs? *The amount of air exchanged*
      after a forced inspiration and a forced expiration.

(e) What is *residual air*? *The amount of air remaining in the*
      lungs after a forced expiration.

(f) Name the process by which carbon dioxide from the blood is transferred to the air
   in the lungs. *Diffusion*

(g) In which structures in the lungs does the transfer of carbon dioxide take place?
      Alveoli

(UCLES)

10

	A Quiet inspiration		**B** Inspiration during strenuous exercise	
	(a) volume/cm^3	(b) number/minute	(a) volume/cm^3	(b) number/minute
Reading (i)	491	10	2400	21
(ii)	505	11	2340	20
(iii)	480	11	2430	19
(iv)	495	12	2390	20
(v)	510	11	2380	20
(vi)	519	11	2400	20
Mean	500	11	2390	20

The table shows the volume of air inspired, and the number of breaths per minute, of a student, **A** resting, **B** undertaking strenuous exercise.

(a) What is the average tidal volume for this student? .... *500 cm^3* .... **(1)**

(b) What is the average volume of air taken into the lungs per minute during quiet inspiration? *500 x 11 cm^3 = 5500 cm^3 min^{-1}* .... **(2)**

(c) Suggest two reasons why there is a range of between 480 cm^3 and 519 cm^3 in quiet inspiration.

(i) *Other external activities requiring oxygen, e.g. talking* .... **(1)**

(ii) *Other internal activities requiring oxygen, e.g. heartbeat, gut movement* .... **(1)**

(d) Give two other important physiological changes you would expect to occur in the body during strenuous exercise.

(i) *Increase in carbon dioxide levels in the blood* ... **(1)**

(ii) *Decrease in muscle glycogen OR increase in lactic acid* .... **(1)**

(e) What would you expect to happen to the volume and rate of breathing in **B** as soon as the strenuous exercise stopped?

Inspirational volume would decrease and the number of inspirations would also decrease (return to normal) .... **(1)**

11 The graph shows the changes in the volume of air in a person's lungs during two minutes.

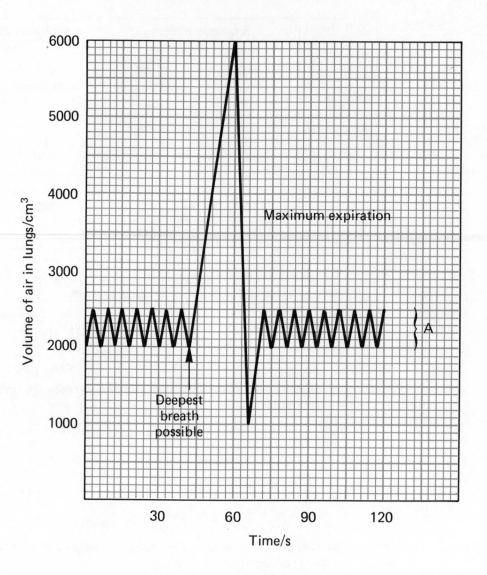

(a) How many times did the person breathe in during the first minute?

Answer ....

(b) '**A**' is called the tidal volume. What is its value for this person?

Answer ....

(c) What do we call the greatest volume of air (maximum expiration) that can be expired after the largest possible breath in?

....

(d) What is its value on the graph?

Answer ....

(e) What is the residual volume of air in the lungs of this person?

Answer ....

(f) What would you expect to happen to the breathing rate during a period of great activity?

....

(g) Explain why this change occurs. ....

....

....

(8)

(h) The table shows some results obtained in an experiment when a person breathed in air containing different amounts of carbon dioxide.
Study the table and then answer the questions.

% Carbon dioxide in inspired air	0.04	0.8	1.50	2.30	3.10	6.00
Tidal volume (cm³)	520	740	790	910	1230	2100
Breathing rate (breaths per minute)	14	14	15	15	16	28

(i) What is the usual percentage of carbon dioxide in the air?   *Answer.* ....

(ii) At 6% carbon dioxide in the air

   (1) What is the tidal volume?   *Answer.* ....

   (2) What is the breathing rate?   *Answer.* ....

(iii) At 0.04% carbon dioxide in the air

   (1) What is the tidal volume?   *Answer.* ....

   (2) What is the breathing rate?   *Answer.* ....

(iv) What effects does an increase in the percentage of carbon dioxide have on inspiration?

....

....

(v) What is the volume of air exchanged in one minute at 1.5% carbon dioxide? Show your working.

....

....

....

(8)
(LREB)

Free-response Question

(a) Explain how you get fresh air into your lungs.   **(9)**

(b) Use a drawing to help you to explain how oxygen from the air in an alveolus reaches haemoglobin in the blood.   **(8)**

(c) How are germs and dust removed from the lungs?   **(3)**

(UCLES)

Answer

(a) The muscle of the diaphragm contracts and the diaphragm flattens from its arched position. The external intercostal muscles contract and the ribs move outwards and upwards. As a result of these two actions, the volume of the thorax is increased. Since the thorax is an airtight chamber, the internal pressure is reduced as a consequence. The lungs follow the outward movement of the thoracic wall, so that their volume and pressure are equally affected.

115

Atmospheric pressure now forces air down the trachea, bronchi and bronchioles and into the alveoli, which are dilated.

(b)

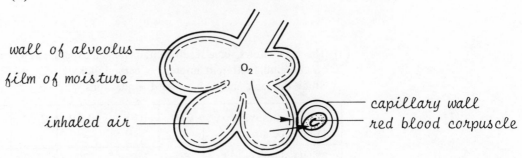

Inhaled air contains 20% of oxygen and there is a much lower concentration of this gas in the capillaries of the alveoli. The oxygen dissolves in the surface fluid of the alveolar walls. Owing to the concentration gradient, the oxygen now diffuses through the alveolar and capillary walls. Both of these structures are only one cell in thickness. On reaching the blood in the capillary, the oxygen enters the plasma in solution. The haemoglobin in the red blood cells has a great affinity for oxygen and so the two combine to form oxyhaemoglobin.

(c) The living cells of the walls of the trachea and bronchi produce mucus which traps germs and dust. Ciliated cells in the epithelium beat towards the larynx, and therefore the mucus and its contents is carried upwards. This eventually passes through the larynx and reaches the throat, where it is swallowed.

Notes

1. Section (a) carries most marks and requires the complete story of inhalation in terms of muscles, volume and pressure changes. See Table 7.1.
2. The drawing is not optional and is explicitly requested; nevertheless it probably does not carry more than two or three marks. It must therefore be kept very simple and take little time to draw. The adjacent walls of alveolus and capillary must be shown together with a red blood corpuscle.

(c) Answers to Objective and Structured Questions

(i) *Multiple-choice Questions*

1. C   2. B   3. B   4. C   5. A   6. C   7. C   8. C   9. D   10. D   11. D   12. C
13. C   14. C   15. B   16. A   17. A

(ii) *Structured Questions*

1 (a) oxygen + sugar → carbon dioxide + water + energy

OR: $6O_2 + C_6H_{12}O_6 \rightarrow 6CO_2 + 6H_2O + 2.81 \times 10^6$ J

Respiration uses oxygen and produces carbon dioxide.
Photosynthesis uses carbon dioxide and produces oxygen.

(b) Ribs	move up	move down
Diaphragm	flattens	arches
Lungs	increase in volume	decrease in volume

(c)

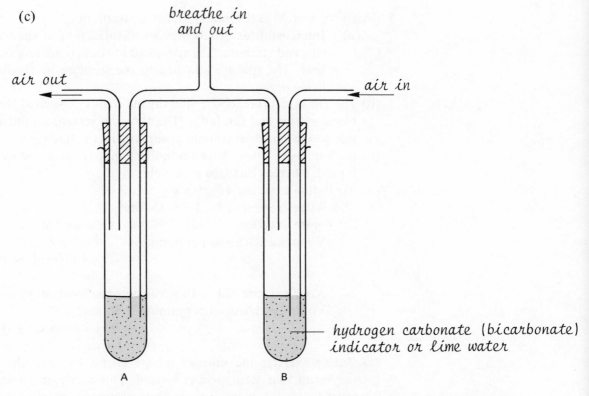

breathe in and out

air out

air in

hydrogen carbonate (bicarbonate) indicator or lime water

A   B

Procedure: Breathe in and out through the mouthpiece. Air bubbles in through test-tube B and out through test-tube A.

Expected results: The contents of tube A turn yellow OR turn milky. Tube B remains red or clear.

Conclusion: There is more carbon dioxide in exhaled air than in inhaled air.

2 (a) 27%   (b) (i) 49%   (ii) 56%   (c) (i) 20%   (ii) 17%
  (d) Emphysema or fatty diet or chronic bronchitis.
  (e) Aging of the body cells and increased susceptibility to infection.
  (f) 1 Nicotine in the maternal blood passes to the foetus.
      2 The foetus is underweight at birth.

3 (a) **A**, larynx; **B**, trachea; **C**, bronchiole; **D**, diaphragm; **E**, superior vena cava; **F**, aorta; **G**, pulmonary artery; **H**, right atrium; **I**, right ventricle; **J**, left ventricle
  (b) (i) Intercostal muscles   (ii) Flattens when breathing in, arches upwards when breathing out.
  (c) Right atrium, right ventricle, pulmonary artery, lungs, pulmonary vein, left atrium.
  (d) Red blood cells; white blood cells.
  (e) Glucose; urea.

4 Conclusion	Column 1	Column 2
(a)	S	No shortage of oxygen and the dotted line rises rapidly
(b)	N	
(c)	C	6% carbon dioxide line rises; 100% oxygen line falls
(d)	N	
(e)	S	rate increases continuously from 0% to 0.03% to 6% carbon dioxide
(f)	C	Increased rate has 94% oxygen compared with only 21% oxygen in normal air

5 (a) (i) **A**, vertebrae (thoracic); **B**, rib; **C**, sternum.
   (ii) Intercostal (external) muscles. Contraction of these muscles raises the ribs and sternum. The ribs pivot at the articulating points on the vertebrae. The ribs are attached to the sternum by flexible cartilage in the living organism.

(b) The rib cage protects the vital organs enclosed, such as the heart, the major blood vessels and the lungs. The thoracic vertebrae contribute towards the role of the vertebral column in supporting the body.

(c) Both of these tissues have a non-living matrix produced by living cells. Bone is rigid, whereas cartilage is flexible.

(d) (i) Before exercise: $15 \times 0.5 = 7.5$ dm^3
       After exercise: $24 \times 1.5 = 36$ dm^3
   (ii) Before exercise: $21 - 17 = 4\%$ oxygen absorbed
       Volume exchanged per minute $= 7.5$ dm$^3 \times 4\%$
       $= 7.5 \times 4$ divided by 100
       $= \underline{0.3 \text{ dm}^3}$
       After exercise: $21 - 16 = 5\%$ oxygen absorbed
       Volume exchanged per minute $= 36$ dm$^3 \times 5\%$
       $= 36 \times 5$ divided by 100
       $= \underline{1.8 \text{ dm}^3}$

(e) During exercise the muscles release energy by anaerobic respiration, with the result that lactic acid is formed. This results in an oxygen debt which must be repaid by an increased rate of oxygen uptake after exercise. At the same time any carbon dioxide which has built up is eliminated by this increased rate.

6 (a) 18%   (b) 40%   (c) 70%   (d) Tar   (e) Emphysema and bronchitis

7 (a) Alveolar air   (b) Atmospheric air   (c) 1.4%

(d) Not all air in the lungs is exhaled. Residual air left in the alveoli collects a higher carbon dioxide content.

(e) Atmospheric air   (f) Alveolar air   (g) 7.1%   (h) (i) Blood entering the lungs   (ii) Blood leaving the lungs   (j) 8.0 cm^3

(k) Produced during respiration of the body cells and carried to the lungs by the blood.   (l) 8.4 cm^3 (m) Oxygen is absorbed from the air in the alveoli.

(n) The diaphragm muscle contracts, pulling down the tendinous sheet of the diaphragm. The intercostal muscles contract, lifting the rib cage. The volume of the thorax is thus enlarged and the internal pressure reduced. Atmospheric pressure therefore forces air into the lungs. When the diaphragm and intercostal muscles relax, the diaphragm arches and the thoracic cage is lowered. The internal volume of the thorax is thus decreased and the increased pressure forces air out of the lungs. The lungs also return to their smaller size, owing to their own elasticity.

Questions **8–10** have the answers supplied with the questions. Question **11** has no answers supplied. Try completing this question yourself.

8 Homeostasis

8.1 Homeostasis

Homeostasis is the maintenance of a constant internal environment (steady state) in animals, e.g. in Man a constant temperature, osmotic pressure of body fluids, pH and concentrations of dissolved substances such as glucose and carbon dioxide.

Three groups of variables needing internal adjustments are:

1. The **intake of food, water and oxygen** required for growth, repair and respiration (energy production) immediately affect the balance of substances within the body.
2. **Waste materials** produced by metabolic activity change the levels of certain substances and these must be controlled, with a return to the steady state.
3. **External environmental factors** also exert influences, e.g. heat, cold, humidity. As a result the body must again make adjustments.

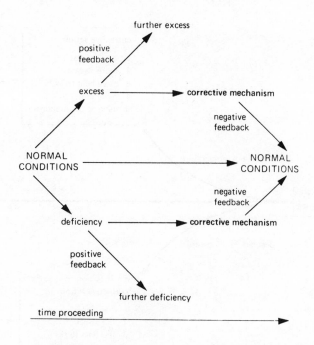

Fig. 8.1 Homeostasis in its simplest form

Man functions within narrow physiological limits: e.g. 'normal' body temperature range is quite small (see Fig. 8.2). In this respect, consider two steady state mechanisms: (1) temperature and (2) water and ionic control (excretion).

8.2 Temperature

See Figs. 8.2 and 8.3 and Table 8.1.

119

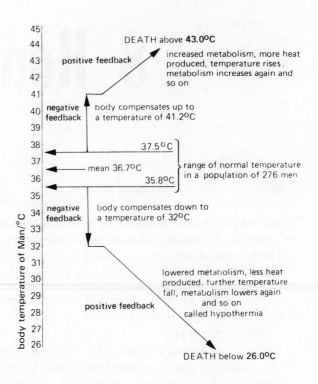

Fig. 8.2 Normal and abnormal temperature range in Man

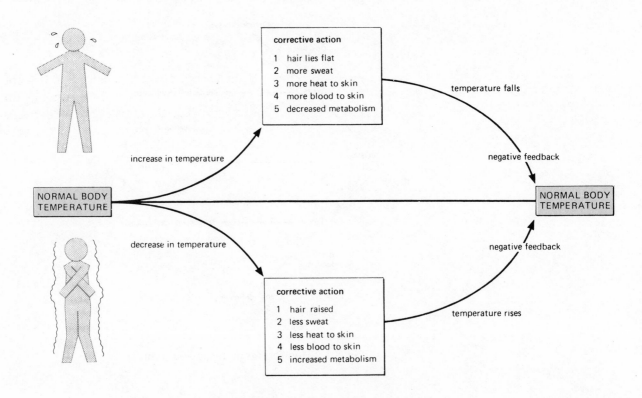

Fig. 8.3 Homeostatic temperature control

Man is **homeothermic**, i.e. he maintains a steady body temperature (like other mammals and birds), as distinct from **poikilothermic** (all other animals, which have a variable temperature directly related to environmental temperatures).

Body temperature in Man is controlled by the hypothalamus (see Section 11.3). Very slight rise in temperature results in the hypothalamus discharging nerve

Table 8.1   Methods by which Man loses and gains heat

Methods by which Man loses heat	*Methods by which Man gains heat*
1. Production of sweat increases. The water in the sweat evaporates, drawing latent heat from the body	Production of sweat decreases, so that heat lost by evaporation is much less
2. Arterioles relax (**vasodilation**) and more blood enters the capillary network. Extra heat is lost by radiation and convection from the skin	Arterioles constrict (**vasoconstriction**) and less blood enters the capillary network. Less heat is lost by radiation and convection
3. The rate of metabolism decreases, so that less heat is produced; thus Man's activity is less in hot weather	The rate of metabolism increases, producing more heat. Loss of heat can cause involuntary muscular action called shivering
4. Behavioural methods: wearing fewer clothes, taking cold baths or swimming, drinking cold drinks, using a fan or air conditioning	Behavioural methods: wearing more clothes, drinking hot drinks or eating hot food, heating houses, taking exercise
5. The hair is lowered in other mammals, making a thinner coat so that the heat can escape more easily. Man can get no benefit from this, since his skin is largely naked	The hair is raised in other mammals, making a thicker coat, thus trapping more air as an insulating layer. In Man the naked skin with its few hairs shows traces of this function. The contracted hair muscles appear as 'goose pimples'

impulses that set in action mechanisms to cool the body. Examine Fig. 8.4 and note that swallowing ice actually raises the temperature of the skin, although the internal body temperature falls, while at the same time the sweating rate decreases. These facts show that temperature control of sweating is not situated in the skin.

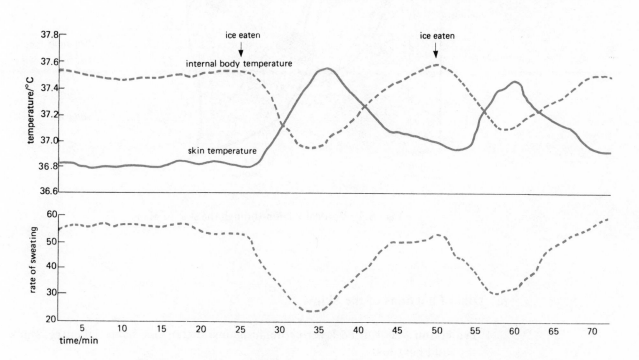

Fig. 8.4   Graphs of internal body temperature and rate of sweating of a man in a temperature controlled at 45°C

The internal body temperatures were taken at the eardrum, showing that the drop in temperature due to eating ice is quickly conveyed by the blood to the brain (hypothalamus).

Heat gain and heat loss (see Table 8.1) are through the relatively hairless skin of Man by radiation, convection and conduction under conditions of low temperature. Heat is lost rapidly from fingers, toes, ears and nose, i.e. body extremities. Too much heat loss over a long period results in a cut-off of blood supply, which can produce 'frost-bite'. The tissues can die through lack of oxygen, food and heat (see Fig. 8.5). Babies lose heat quickly because of their low surface area-to-volume ratio. They move little, so do not generate heat. In temperate and cold climates they should sleep in a warm bedroom and be well covered with bed clothes. Old people suffer the same problems, often dying through extreme heat loss, hypothermia.

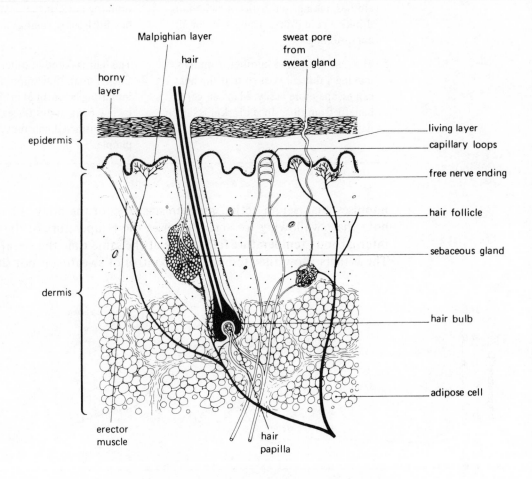

Fig. 8.5 Vertical section through the skin of Man

(a) Other Functions of the Skin

1. **Protection** — cornified layer protects against water loss, bacterial entry, sun's rays and heat loss.
2. **Vitamin D production** in sunlight.
3. **Energy storage** in the form of fat in the lower layers of the dermis.

(b) Skin Hygiene

Sebum and sweat on the skin are an ideal breeding ground for bacteria. Therefore, infection will result if the skin is not kept clean. Regular washing with soap and water removes natural oils and sweat. Hands should be cleansed after visits to the lavatory.

8.3 Excretion

Excretion and osmoregulation are the main functions of the kidneys. For the external and internal structure of the kidneys and their associated organs, see Figs. 8.6 and 8.7.

The kidney is a mass of kidney tubules or nephrons.

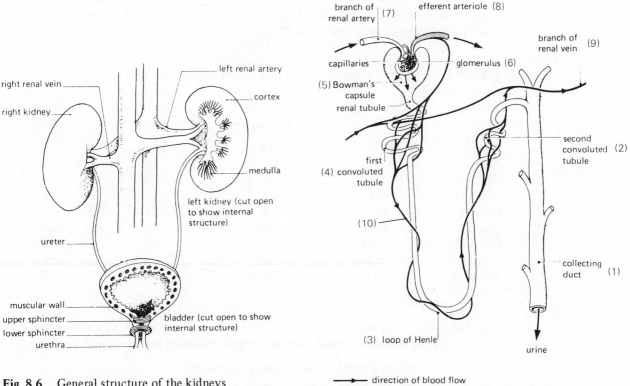

Fig. 8.6 General structure of the kidneys and bladder in Man

⟶ direction of blood flow

Fig. 8.7 Diagram of one renal tubule, showing the blood supply and the main areas of reabsorption (numbers refer to questions 6–12 in the multiple choice questions)

(a) The Work of the Nephron

The breaking-down activities of living cells (catabolism) result in excretory products, the most abundant of which are water, carbon dioxide and ammonia. These are dangerous (toxic), since they may accumulate in excess of the needs of the body. **Excretion is therefore the elimination of the toxic products of metabolism.**

1. Blood enters the capillaries of the glomerulus from the renal arteries under normal blood pressure. This blood pressure is increased because the exit capillary is **narrower** in bore than the **capillary delivering blood** (see Fig. 8.7).

123

2. Blood under high pressure is forced through the capillary wall into Bowman's capsule. This process is called **ultrafiltration**, because it is brought about by the enhanced blood pressure.
3. Small molecules, namely water, salts, urea, uric acid, glucose and amino acids, are forced through, whereas red blood cells and large molecules (proteins, e.g. fibrinogen) remain in the circulating blood.
4. The rate of production of tubule fluid is about 130 cm^3 per minute (about 100 dm^3 per day). The final production of urine is about 1500 cm^3 per day. Thus, about 98% of the fluid passed into the tubules is reabsorbed.
5. The fluid passes down the nephron and the following occurs:

   (a) Glucose, amino acids, salts and most water are reabsorbed.
   (b) These substances are forced back into the tubule capillaries **against the diffusion gradient**, so that **energy is required**. Thus, increased **respiration** in the **mitochondria** of the cells in the tubule walls is essential.
   (c) After this **selective reabsorption** has occurred, the liquid in the nephron consists of urea, uric acid, salts, water and other nitrogenous waste materials (see Table 8.2).

Table 8.2 Concentrations of substances in blood plasma and urine

Substance	% in plasma	% filtrate into nephron	% in urine	Concentration factor
Water	90–93	90–93	95.0	
Protein	7.0	0	0	
Glucose	0.1	0.1	0	
Sodium	0.3	0.3	0.35	× 1.0
Chloride	0.4	0.4	0.6	× 1.5
Urea	0.03	0.03	2.0	× 60.0
Uric acid	0.004	0.004	0.05	× 12.0
Creatinine	0.001	0.001	0.075	× 75.0
Ammonia	0.001	0.001	0.04	× 40.0

   (d) Some salts and water can be reabsorbed in varying quantities according to the amount of each substance in the blood (see section 8.1):

   dilute plasma — more water eliminated
   concentrated plasma — more water reabsorbed

   (e) Remaining urine trickles into the collecting duct and thence flows through ureters to the bladder.

6. If glucose appears in the urine (i.e. is not totally reabsorbed), owing to excess glucose in the plasma, then the body is suffering from **diabetes**. This is caused by a lack of the hormone **insulin** (see section 8.1).
7. The control of water in the body is the function of the kidneys in conjunction with the **hypothalamus** of the brain, which detects the level of water in the blood. The hypothalamus then causes the pituitary gland to secrete the **anti-diuretic hormone** (ADH), which reduces loss of water in the urine. This process also helps to maintain the osmotic pressure of the blood. If too much water is present in the blood, then the secretion of ADH decreases (see Fig. 8.8).
8. The bladder has elastic tissues which enable it to expand to 500 cm^3. The pressure of the urine triggers a reflex and impulses are sent to the bladder from the spinal cord. The sphincter muscle at the exit of the bladder relaxes

Table 8.3 Summary of substances excreted from body organs

Lungs	Skin	Kidneys
Water	Water	Water
Carbon dioxide	Sodium chloride	Sodium chloride
	Urea	Urea
		Uric acid
		Creatinine
		Ammonia

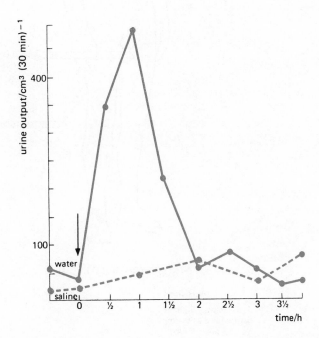

Fig. 8.8 Graph of the response to drinking 1 litre of water and, on the following day, 1 litre of 0.96% sodium chloride solution by a healthy man

and with contraction of bladder muscles the urine is expelled down the urethra.

(b) Artificial Kidney

In order to relieve the kidney of its work during certain diseases, the patient is often connected to an artificial kidney machine. The patient's blood is chanelled off through a long coiled tube. The tube is immersed in a prepared solution of salts of approximately the same concentration as blood. The tube is composed of a membrane permeable to urea, uric acid, sodium, potassium and creatinine, substances normally excreted by the kidney tubules. Blood cells and plasma proteins do not cross the membrane. The result is that the blood returns to the body cleansed of its excretory substances. The passage of these small molecules through the membrane is called **dialysis.**

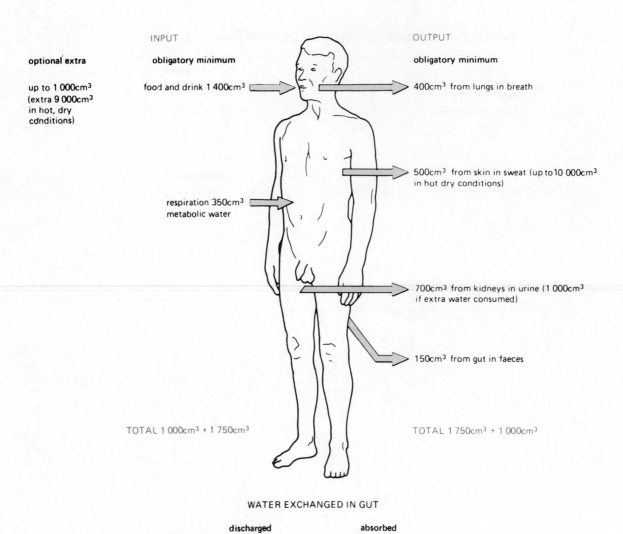

INPUT

optional extra

obligatory minimum

obligatory minimum

up to 1 000cm³
(extra 9 000cm³
in hot, dry
conditions)

food and drink 1 400cm³

400cm³ from lungs in breath

500cm³ from skin in sweat (up to 10 000cm³
in hot dry conditions)

respiration 350cm³
metabolic water

700cm³ from kidneys in urine (1 000cm³
if extra water consumed)

150cm³ from gut in faeces

TOTAL 1 000cm³ + 1 750cm³

TOTAL 1 750cm³ + 1 000cm³

WATER EXCHANGED IN GUT

	discharged	absorbed
saliva	1 500cm³	
gastric juice	2 500cm³	
bile	500cm³	
pancreatic juice	700cm³	
intestinal juice	3 000cm³	
	8 200cm³	8 200cm³

Fig. 8.9 Water gain and water loss in Man

8.4 Questions and answers

(a) Multiple-choice Questions

1 Which of the following is the best definition of homeostasis?
   A control of temperature
   B control of water and ionic levels
   C independence of the external environment
   D maintenance of a constant internal environment
   E maintenance of constant body functions by the external environment

2 Which of the following is a physiological method of cooling the body?
   A sitting in an air-conditioned room
   B taking a cold shower or bath
   C producing more sweat on the skin surface
   D drinking a hot drink
   E eating less food

126

3 Which of the following lose heat at the fastest rate?
   A tall, thin men in the tropics
   B tall, thin men in the Arctic
   C short, fat men in the tropics
   D short, fat men in the Arctic
   E short, thin men in the tropics

4 An increase in the carbon dioxide concentration in the blood of a man results in
   A negative feedback and a slower breathing rate.
   B positive feedback and a slower breathing rate.
   C no feedback and normal breathing rate.
   D positive feedback and a faster breathing rate.
   E negative feedback and a faster breathing rate.

5 Which one of the following aids heat loss from the skin of a man on a hot, dry day?
   A subdermal fat
   B dermal blood vessels
   C light clothing
   D hair
   E sebaceous glands

Examine Fig. 8.7 and answer questions 6–12

6 Into which one of the following structures does ultrafiltration of blood occur?
   A 2
   B 3
   C 4
   D 5
   E 9

7 Which one of the following structures directly conveys urine to the ureters?
   A 1
   B 2
   C 4
   D 5
   E 6

8 From which of the following is most water absorbed back into the bloodstream?
   A 2
   B 3
   C 4
   D 5
   E 6

9 Glucose molecules in the filtrate are quickly reabsorbed into the blood. Which one of the following processes must occur before reabsorption can take place?
   A Glucose molecules are ionised.
   B Water is lost rapidly.
   C Respiration supplies energy.
   D Tubule membranes become impermeable.
   E More urea is absorbed to balance the loss of glucose.

10 Which one of the following blood vessels contains the highest concentration of urea?
   A 6
   B 7
   C 8
   D 9
   E 10

11 Which one of the following blood vessels contains the highest concentration of glucose?
   A 6
   B 7
   C 8
   D 9
   E 10

12 Which one of the following substances is at its highest concentration in 1 compared with its concentration in 5?

A uric acid

B ammonia

C urea

D creatinine

E glucose

13 Which one of the following is an excretory function of the liver?

A conversion of glycogen to glucose

B conversion of amino acids to urea

C conversion of fats to fatty acids and glycerol

D production of vitamin B_{12}

E metabolism of glucose to glycogen

14 Which of the following is found in both respiratory and excretory organs?

A waste removal from a large surface area

B secretion of hormones to produce homeostasis

C excretion of waste nitrogenous substances

D control of the osmotic concentration of the blood

E regulation of carbon dioxide levels in the blood

15 The kidney is a homeostatic organ because it

A removes digested food from the body.

B prevents loss of urea from the blood.

C keeps tissue fluid at a constant level.

D regulates the ionic content of the blood.

E removes sugar and excretes it in the urine.

16 Which of the following is the name given to the minute tubules of which the kidney is mainly composed?

A dendrons

B neurons

C axons

D nephrons

E neutrons

Table 8.4   Measurements on one man

	8 Jan.	25 Mar.	10 July	16 Sept.	20 Oct.
Mouth temperature (°C)	39.1	36.8	37.1	36.9	36.6
Breathing rate per minute	28	16	18	30	15
Pulse rate per minute	100	70	85	120	75
% sugar concentration in urine	0.02	4.2	0	0	0

Table 8.4 shows five groups of data obtained on different days throughout a period of one year by measurements on one man. Examine the data and answer questions 17–20.

17 On which one of the following dates had he run up a flight of stairs immediately before measurements were made?

A 8 January

B 25 March

C 10 July

D 16 September

E 20 October

18 On which one of the following dates was the man suffering from influenza?

A 8 January

B 25 March

C 10 July

D 16 September

E 20 October

19 On which one of the following dates was it discovered that the man was suffering from diabetes?
   A 8 January
   B 25 March
   C 10 July
   D 16 September
   E 20 October

20 On which one of the following dates was it clear that he had been treated and was containing the diabetes?
   A 8 January
   B 25 March
   C 10 July
   D 16 September
   E 20 October

21 Which one of the following is the general definition of excretion?
   A removal of urine by the bladder
   B removal of faeces from the gut
   C removal of urea from the liver
   D removal of water by the skin
   E removal of waste products of metabolism

22 Which one of the following is formed from nitrogen compounds by the liver?
   A urine
   B amino acids
   C nitrates
   D urea
   E proteases

23 Which one of the following controls salt levels in the body of humans?
   A rectum
   B bladder
   C lungs
   D kidney
   E liver

24 Which one of the following occurs to the compounds remaining after deamination of amino acids in the liver?
   A converted to carbohydrates
   B used to produce more blood
   C passed out in urine
   D passed out with faeces
   E converted to enzymes

(b) Structured Questions

1 The diagram shows the structure of a nephron (kidney tubule) and its associated blood vessels.

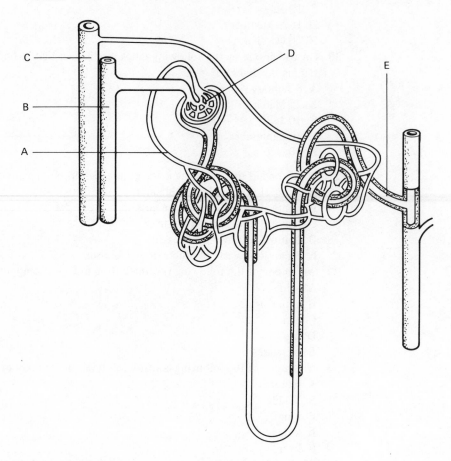

(a) Name the structures labelled **A–D**. (four lines) **(4)**

(b) (i) Name the process which occurs at **D**. (one line) **(1)**

    (ii) What feature of the blood enables this process to take place? (two lines) **(2)**

(c) Name the substance which, apart from water, is present in the nephron and has its highest concentration at point **E**. (one line) **(1)**

(d) Place an arrow and the letter **R** on the diagram to indicate the region where amino acids and glucose are mainly reabsorbed. **(1)**

(e) (i) Use an arrow and the letter **S** on the diagram to indicate the position of the distal convoluted tubule. **(1)**

    (ii) What major constituent of the filtrate, apart from water, is reabsorbed in the distal convoluted tubule? (one line) **(1)**

(L)

2

Substance	1 Amount filtered	2 Amount excreted	3 Amount reabsorbed
Water	180 dm³	1.8 dm³	
Glucose	180 g	nil	
Urea	50 g	39 g	
Sodium	600 g	12 g	
Calcium	5 g	0.2 g	
Potassium	35 g	2 g	

The above table shows daily values for some of the substances filtered from the blood and finally excreted in the urine.

(a) Complete the third column of the table to show the amount of each substance re-absorbed. **(6)**

(b) Which of the substances in the table would be excreted in larger amounts as a result of the following conditions?
  (i) low external (environmental) temperatures (one line) **(1)**
  (ii) consumption of large amounts of meat (one line) **(1)**
  (iii) lack of insulin production (one line) **(1)**

(c) (i) Name the hormone which affects the amount of water reabsorbed in the kidneys. (one line) **(1)**
  (ii) Name the process by which water is reabsorbed in the nephrons. (one line) **(1)**
  (iii) Name the gland which secretes the hormone named in (i) above. (one line) **(1)**

(d) Describe briefly how homeostasis is applicable to kidney function. (three lines) **(3)**

3 (c) The diagram below shows some of the features in the use of amino acids in the body.

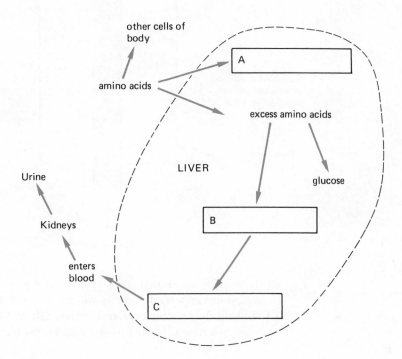

(i) Finish this diagram by putting into the boxes the correct word, selected from the list below:

carbohydrate   ammonia
protein   carbon dioxide
pancreas   urea **(3)**

(ii) Why are more amino acids taken into the body than are needed for the normal functioning of the body? (three lines) **(2)**

(iii) Name *two* blood vessels which substance **C** must pass through to reach the kidneys. (one line) **(2)**

(d) Some persons suffer from kidney diseases which lead to the failure of the kidneys to function properly. In each case a kidney machine may be used through which the blood of the patient is passed.

The diagram below represents, in very simplified form, a part of such a kidney machine.

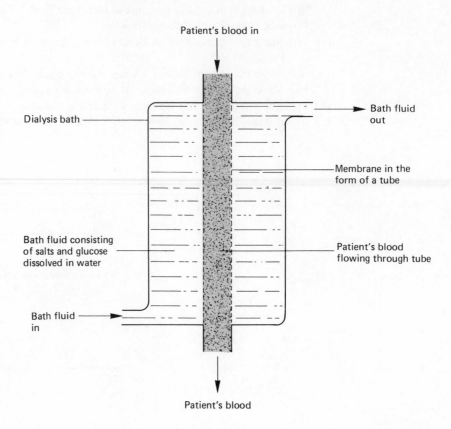

(i) If the patient is not to be harmed in any way by the use of such a machine, suggest one important property which the membrane must have. (one line)   **(1)**

(ii) The bath fluid contains certain dissolved salts in the same concentration as in the blood of a normal healthy person. Suggest the reason for this. (one line)   **(1)**

[Part question] **(SREB)**

4 The figure shows the reproductive system and the urinary system in a human male.

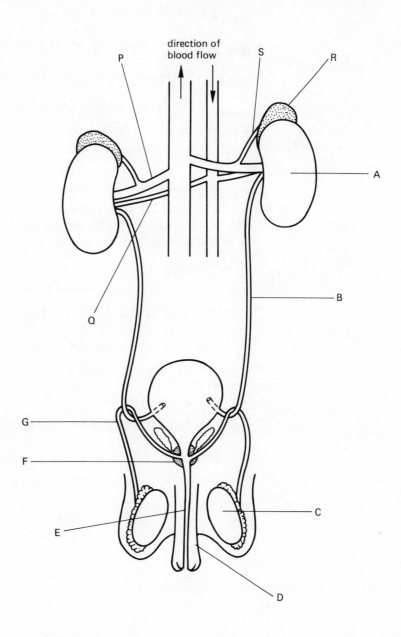

(a) Name the organs labelled.

A..... *Kidney* ....

B..... *Ureter* ....

C..... *Testis* ....

D..... *Penis* .... **(4)**

(b) (i) Which of the vessels **P** or **Q** is the renal **artery**?

.. *Q* .... **(1)**

(ii) State how this can be deduced from the diagram.

..... *Q arises from a narrow vessel bringing blood from the* .....

..... *heart* .... **(2)**

(c) (i) State one function of each of the following.

organ **A** .*Production of urine*......

gland **F** .*Production of seminal fluid*........

tube **B** .*Conduct urine from kidney to bladder*...

tube **G** .*Conduct sperms from testis*........ **(4)**

(ii) State two functions of the tube labelled **E**.

1 .*Pass out urine*........

2 .*Pass out male gametes (sperms)*...... **(2)**

(d) The structure labelled **R** is an adrenal gland.

(i) Name a substance which will leave this gland in blood vessel **S** and which helps to prepare the body in an emergency.

.*Adrenaline*........ **(1)**

(ii) State **two** effects of this substance on the body.

1 .*Makes the heart beat faster.*......

2 .*Increases the rate of breathing.*...... **(2)**

(e) The organ labelled **C** produces sperm cells. These contain only half as many chromosomes as other body cells.

(i) Give **one** explanation for this.

.*During sperm formation cells divide by*......

meiosis, which halves the chromosome number....... **(2)**

(ii) Name the part of the sperm cell which contains the chromosomes.

.*Nucleus*........ **(1)**

(UCLES)

5 (a) The diagram below shows a vertical section of a small piece of skin of a person sitting in a room at 20°C.

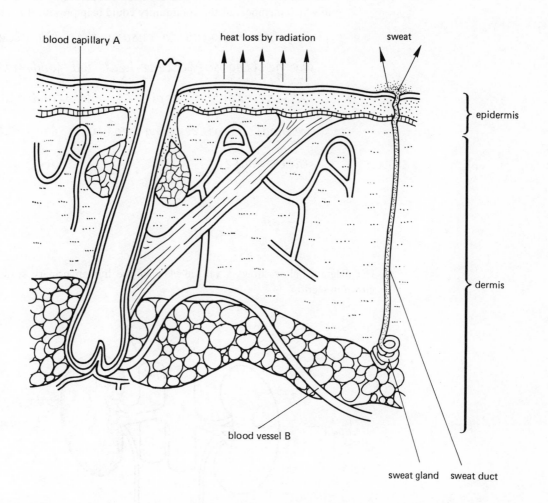

blood capillary A heat loss by radiation sweat

epidermis

dermis

blood vessel B

sweat gland sweat duct

(i) If this person moved to a colder room (10°C), what change would occur to blood vessel **B**?

The vessel would constrict so that its lumen would be smaller. **(1)**

(ii) Explain the change described in (i) above.

The lower temperature would cause more heat loss, but less blood flow to the surface cuts down heat loss. **(1)**

(iii) If this person moved to a colder room (10°C), what change would occur to the sweat gland?

The sweat gland would extract less water and dissolved salts from the blood. **(1)**

(ıv) Explain the change described in (iii) above.

Sweat secreted on evaporation draws heat from the body. Less sweat secreted helps to conserve heat in the body. **(1)**

(v) Every winter some elderly people are found in their homes, dead from hypo-thermia (excessive lowering of the body temperature). Suggest briefly, **two** ways in which members of the community could help prevent this situation.

1. *Regular visiting to ensure that the elderly people are getting sufficient heat and food for survival.*

2. *Assist elderly people to understand and undertake methods of preventing heat loss from themselves and their house.* **(2)**

(YHREB, 1985)

6 The diagram below shows a ventral view of the human urinary system. One kidney is shown in section.

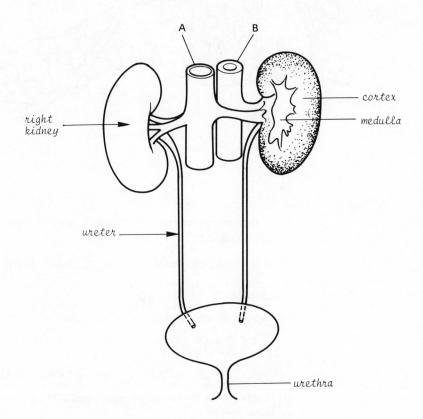

(a) Label, on the diagram, the following structures: right kidney, ureter, urethra, cortex, medulla.

(b) State which blood vessel, **A** or **B**, is the dorsal aorta ...*B*...
Give a reason for your choice.

The transverse section near the label line shows a thicker wall than A; also, the lumen is of a smaller diameter.

(c) The diagram below shows the capsule and capillaries (glomerulus) of a kidney tubule. As the blood flows through the capillaries, certain materials pass into the capsule. These are known as the *filtrate*.

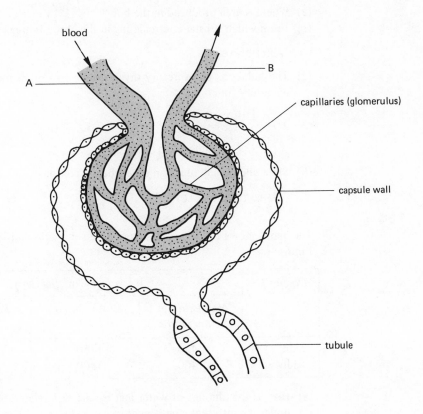

(i) State **one** observable difference between blood vessels **A** and **B**.

Blood vessel B has a smaller diameter than A.

(ii) Explain how this difference helps in the functioning of the capsule.

The blood pressure in the glomerulus is increased, so that substances pass into the capsule by ultra-filtration.

(d) Study the table below, which relates to the functioning of the capsule.

Materials	Concentrations (g/100 cm^3)		
	Blood plasma	Filtrate	Urine
Urea	0.03	0.03	2.00
Glucose	0.10	0.10	0.00
Amino acid	0.05	0.05	0.00
Protein	8.00	0.00	0.00

Which material does **not** pass into the capsule? Explain this.

Material *Protein*

Explanation . . . . . The blood proteins (e.g. gamma globulin) . . . . .

have molecules that are too large to pass through.

(e) Where is urea produced in the body? . .Liver. . . . .

(f) From which substance circulating in the blood is urea produced?

Amino acids
. . . .

(g) The kidney is an excretory organ. Name **two** other mammalian organs that have an excretory function.

(i) . .Lungs. . . . .

(ii) . .Skin (sweat glands). . . . .

(h) State **one** kidney function other than excretion.

Osmoregulation
. . . .

(UCLES)

7 The table shows the water loss from a person's skin and kidneys under different conditions.

Organ	Water lost (cm^3 per day)		
	Normal room temperature	Hot weather	Prolonged heavy exercise
Skin	450	1750	5350
Kidneys	1400	1200	500

(a) How is the amount of water lost by the skin related to the temperature of the body under the different conditions? **(2)**

The higher the body temperature the greater the amount
. . . .

of water lost from the skin.
. . . .

(b) How is the amount of water lost by the kidneys related to the amount of water lost by the skin? **(2)**

The greater the amount of water lost from the skin
. . . .

the less is lost through the kidneys.
. . . .

(SREB)

8 The figure represents part of the urino-genital organs of a human male.

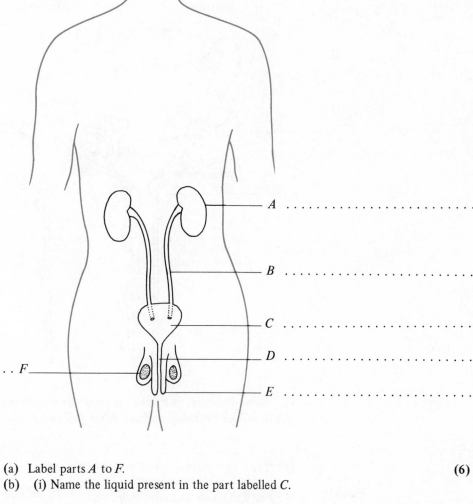

(a) Label parts *A* to *F*.   **(6)**

(b) (i) Name the liquid present in the part labelled *C*.

. . . . **(1)**

(ii) How would you test a sample of this liquid to see if it contained any protein?

. . . .

. . . .

. . . . **(2)**

(iii) What disease could be responsible for the presence of glucose in this liquid?

. . . . **(1)**

(c) The figure below is a diagram of one of the many tubules to be found in the organ labelled *A* in the figure in part (a).

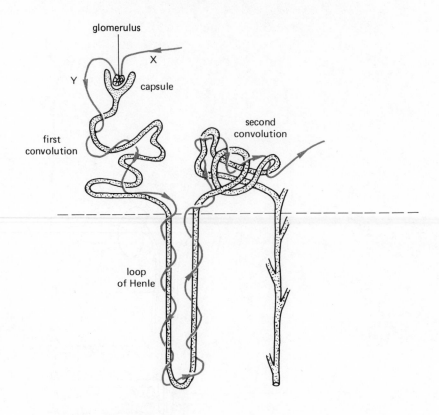

The arrowed line represents the course of the blood vessels around the tubule.
  (i) How does the blood pressure differ at *X* and *Y*?

.... **(1)**
 (ii) Water in the blood at the glomerulus passes into the capsule, taking with it several soluble substances.
Name the main excretory substance.

.... **(1)**
Name **one** other substance.

.... **(1)**
(iii) Name **two** substances reabsorbed into the blood at the first convolution.

....

.... **(2)**
(iv) Why is water reabsorbed into the blood around the loop of Henle?

....

.... **(1)**
(EAREB)

Free-response Question

(a) Define the term 'excretion'. Explain the importance of this process in all living organisms.
(b) What are the functions of the kidney? Describe how the kidneys perform them, and explain why they use up a lot of energy.

Answer

(a) Excretion is the removal from the body of waste products formed as a result of the metabolism of cells.

It is essential for excretion to occur, for the presence of toxic waste products could inhibit the normal functioning of cells in several ways. The waste could alter the osmotic pressure of the cellular environment, could affect the pH or could be poisonous to enzyme systems. Any of these could affect the metabolic activity of cells in a harmful way.

(b) The functions of the kidneys are (i) to remove urea and other waste products (uric acid, sodium chloride) from the blood and then expel it from the body, and (ii) to control the water content of the body (osmoregulation).

Each kidney comprises about one million kidney tubules closely associated with blood capillaries. It is the tubules which perform the actual functions of the kidney.

The capillary entering Bowman's capsule is wider than the one leaving it. This creates a bottleneck in the glomerular capillaries, and, hence, the increased resistance to blood flow causes fluid to be filtered out through the capillary walls and to collect in the cavity of Bowman's capsule. This process is called ultrafiltration.

The fluid contains glucose, amino acids, salts and nitrogenous waste (urea, uric acid and ammonia) dissolved in water. Hence, the blood leaving Bowman's capsule consists mainly of cells, blood proteins and some water. There is, however, still some urea and other substances, for the tubules do not entirely remove all of the excretory products at each pass of the blood.

Clearly, loss of valuable foodstuffs would be disastrous, and thus soluble foods and salts are selectively reabsorbed by the cells of the first coiled tubule. Some water is also reabsorbed. After this, the liquid left in the tubule will contain only waste products such as urea, excess salts and water.

Man cannot afford to lose large amounts of water. The function of the loop of Henle and other parts of the kidney tubule is to remove as much water as possible according to the needs of the body. If the person has drunk a large quantity of water, then the blood will possess a high water content. In order to return it to its normal level, dilute urine will be expelled. If the blood water content is low, then concentrated urine will be excreted.

The second coiled tubule makes the final adjustment to the composition of the blood. Here further useful substances which may have escaped earlier reabsorption are passed back into the blood. Any further uptake of water can also take place here.

Thus, blood leaving the kidney will contain a little urea, will have regained its food and salt content, and will contain an optimum level of water. The urine expelled will be more concentrated than the surrounding tissues as a result of reabsorption of water into the body.

Large amounts of energy are required by the kidney to enable it to function properly, because selective reabsorption of soluble foodstuffs can only occur if there is an energy supply. Similarly, reabsorption of water against the osmotic gradient is an energy-consuming process.

Notes

1. A definition must always be brief.
2. Many candidates confuse defaecation with excretion. The former eliminates mainly undigested food, roughage and bacteria, although the faeces do con-

tain small amounts of excretory matter eliminated in the bile entering the gut from the liver. These are breakdown products of red blood corpuscles.

3. A simple drawing of a nephron (kidney tubule) could be included in this answer to aid the description of its function. It is time-consuming and is not specifically demanded in the question.

4. It is unusual to demand that the candidate consider energy in relation to the work of the kidney. A thoughtful student should be able to answer this section.

(c) Answers to Objective and Structured Questions

(i) *Multiple-choice Questions*

1. D   2. C   3. B   4. E   5. B   6. D   7. A   8. C   9. C   10. B   11. B   12. C
13. B   14. A   15. D   16. D   17. D   18. A   19. B   20. C   21. E   22. D
23. D   24. A

(ii) *Structured Questions*

1 (a) **A**, efferent capillary; **B**, arteriole; **C**, venule; **D**, glomerulus
   (b) (i) Ultrafiltration   (ii) Normal blood pressure is increased as a result of the narrow diameter of the exit vessel of the glomerulus.
   (c) Urea   (d) (Arrow and letter **R**) at the first convoluted tubule
   (e) (i) (Arrow and letter **S**) on tubule just before letter **E**   (ii) Salt

2 (a) Water, 178.2 dm^3; glucose, 180 g; urea, 11 g; sodium, 588 g; calcium, 4.8 g; potassium, 33 g
   (b) (i) Water   (ii) Urea   (iii) Glucose
   (c) (i) ADH, or antidiuretic hormone (vasopressin)   (ii) Osmosis
        (iii) Pituitary gland
   (d) The regulation of water balance in the body is achieved by the effect of a hormone (ADH) acting on the distal convoluted tubule and collecting ducts. Increased water results in a fall of plasma osmotic pressure, decrease in ADH and copious clear urine produced. Decreased water results in a rise of plasma osmotic pressure, increase in ADH and a reduced volume of concentrated urine production.

3 (c) (i) **A**, protein; **B**, ammonia; **C**, urea
        (ii) They are derived from protein that is taken in during feeding. In general, more protein is consumed than is required for growth and maintenance.
        (iii) Any two of the following: hepatic portal vein/posterior vena cava/ pulmonary artery/pulmonary vein/aorta/renal artery
   (d) (i) Sterile/non-toxic/non-degradable
        (ii) Bath fluid and blood have the same concentrations of salts so that patients' blood remains stable. Only excretory molecules with higher concentration in the blood will pass out into the bath fluid.

Questions **4–7** have the answers supplied with the questions. Question **8** has no answers supplied. Try completing this question yourself.

9 Support, Muscles and Locomotion

9.1 Essentials of a locomotory system

1. A **contractile tissue** (muscle) acting upon
2. a **rigid skeleton**, which in turn acts upon
3. a **resisting medium** (earth when walking, water when swimming).

(a) Functions of the Skeleton

1. To **support** the body. (The surrounding air offers no support.)
2. **Protection** — a bony box (cranium) surrounds the brain; the rib cage surrounds the heart and the lungs; the vertebral column surrounds the spinal cord.
3. **Locomotion** — the action of muscles on bones to move the limbs. The bones are jointed, to give a lever action.
4. Non-structural functions: (a) production of **red blood corpuscles** and **white blood cells**; (b) a **reserve** of calcium and phosphorus.

9.2 Tissues of the skeleton

1. **Bone** — hardness is caused by calcium salts (mainly calcium sulphate) which form part of a matrix (70% bone). The matrix is secreted by bone cells and perforated by Haversian canals. The canals contain blood vessels. Bone-secreting cells form concentric circles around the canals. Bone marrow lies in the central cavity of the bone and continues into the bone, giving a spongy appearance.
2. **Cartilage** (or gristle) is present at the ends of long bones and has a clear smooth structure. It is elastic in nature, owing to protein fibres secreted by its cells. It acts as a shock absorber and reduces friction at the joints.
3. **Ligaments** consist of tough, fibrous, elastic tissue protecting joints.
4. **Tendons** consist of tough, fibrous, inelastic tissue attaching muscle to bone. They must be inelastic to enable the muscle contraction to be conveyed to bone.

143

9.3 The skeleton

(a) The Axial Skeleton

The **axial** skeleton (see Figs. 9.1 and 9.2) is made up of the skull and the vertebral column. The skull consists of the brain-box (**cranium**) and the fused upper jaw together with the lower jaw. The cranium is formed of bony plates joined together. In a new-born baby the bones have not fused at the top of the cranium and this results in delicate, unprotected areas, the **fontanelles.**

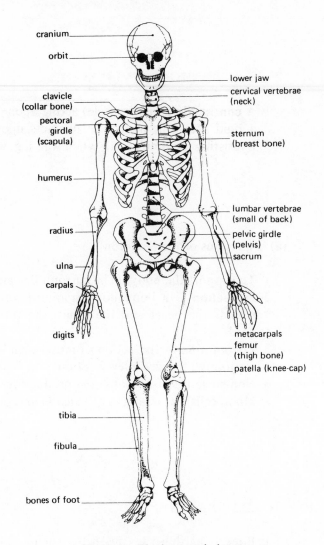

cranium
orbit
lower jaw
cervical vertebrae (neck)
clavicle (collar bone)
pectoral girdle (scapula)
sternum (breast bone)
humerus
lumbar vertebrae (small of back)
radius
pelvic girdle (pelvis)
sacrum
ulna
carpals
digits
metacarpals
femur (thigh bone)
patella (knee-cap)
tibia
fibula
bones of foot

Fig. 9.1 The human skeleton

The **vertebral column** is made up of 34 bones called vertebrae. Each succeeding pair of vertebrae is separated by a pad of cartilage termed the **intervertebral disc**. Twenty-eight of the vertebrae are moveable and flexible. Each has a central disc of bone, the **centrum**, to the dorsal surface of which is attached the **neural arch**. Within successive neural arches lies the **spinal cord** within the **neural canal**. Seven bony projections may arise from the neural arch for attachment of muscles and articulation between vertebrae. Two **transverse processes** arise from the junction of the neural arch with the centrum, and the **neural spine** arises centrally from the neural arch. Four articulating **facets**, two anterior and two posterior, are present. The downward-facing facets on the posterior end of the vertebra fit into the upward-facing facets of the vertebra immediately behind (see Figs. 9.2 and 9.3).

The seven **cervical** vertebrae support the neck and the head. The first is the **atlas**, permitting the head to rotate, and the second is the **axis**, permitting nodding of the head. The twelve **thoracic** vertebrae articulate with the ribs, which can move up and down on the vertebrae during breathing. The **ribs** are joined to the **sternum** by flexible cartilage. The five **lumbar** vertebrae are large and weight-bearing with large processes for muscle attachment. The five **sacral** vertebrae are fused and attached to the pelvic girdle. Five small fused vertebrae form the **coccyx**, the remnants of a tail.

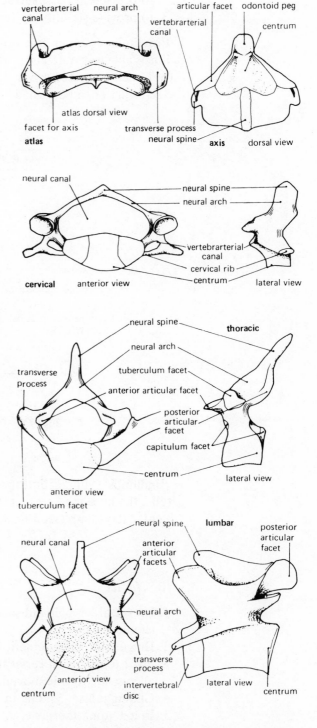

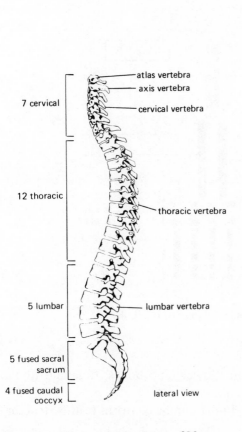

Fig. 9.2 The vertebral column of Man

Fig. 9.3 The different types of vertebrae in Man

(b) The Appendicular Skeleton

See Figs. 9.1 and 9.4. The **pectoral girdle** consists of two **scapulae** (shoulder-blades) and two **clavicles** (collar-bones). The girdle is incomplete, as the two scapulae do not join but move freely over the shoulders. The **pelvic girdle** is a complete ring of bone fused to the **sacrum** for strength. It is wider in women to permit childbirth and is enabled to enlarge by a flexible cartilage between the pubic bones.

The girdles have three main functions:

1. to form a more or less rigid connection between axial skeleton and limbs;
2. to provide suitable surfaces for attachment of muscles that move limbs;
3. to provide stability by separating the limbs.

Forelimbs — upper arm bone, humerus; lower arm bones, radius and ulna. The radius is able to rotate about the ulna (see Fig. 9.4). The opposable thumb enables a tool, such as a screwdriver, to be grasped, and the rotation of the forearm allows the screwdriver to be turned.

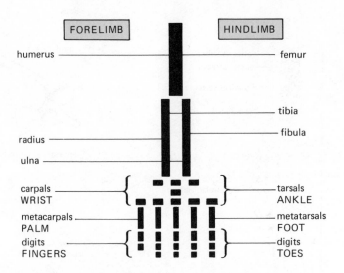

Fig. 9.4   The pentadactyl plan of the limbs

Hindlimbs — upper limb bone, femur; lower limb bones, tibia and fibula. The patella, the knee-cap, protects the knee joint.

Pentadactyl limb (see Fig. 9.4). The human limbs conform to this structure.

(i) *Joints*

1. Universal or **ball-and-socket** joint — movement in any plane, e.g. hip and shoulder.
2. **Hinge** joint — movement in one plane, e.g. knee and elbow.
3. **Pivot** joint — first cervical (atlas) pivots around the second cervical (axis) and allows side-to-side movement.
4. **Gliding** joint — gliding movement, e.g. ankle and wrist.
5. **Fixed** joint (suture) — no movement, e.g. between bones of the skull or pelvic girdle.

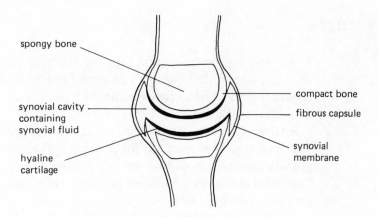

Fig. 9.5 A synovial joint

9.4 Muscles and movement

Muscles can only contract and relax (they are unable to **increase in length**). Therefore, they must work in pairs to effect movement, with one muscle lengthened by another pulling it. These are called **antagonistic pairs**, e.g. biceps and triceps moving the lower arm (see Fig. 9.11(c)). Muscles that bend limbs are called **flexors**, while those that straighten limbs are **extensors**.

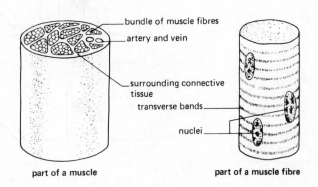

Fig. 9.6 Voluntary muscle

Skeletal muscle (striped, striated or voluntary muscle) is formed of fibres each of which has several nuclei and is distinguished by a series of transverse bands parallel to its short axis. The fibres form bundles, and a muscle is made up of a number of bundles, well supplied with blood vessels and nerves.

Energy for contraction comes from **glycogen** in the muscle, which is oxidised to carbon dioxide and water, releasing energy. This becomes bound up in **ATP molecules** (see Chapter 7). Skeletal muscle working unceasingly becomes **fatigued**, but given a period of rest the muscle recovers and can work again. The fatigue is due to the build-up of **lactic acid**, and with rest this is oxidised.

Muscle cramp is due to lack of nutrients or oxygen coming to the muscle or to an accumulation of waste products. The reason is often an interference of blood flow caused by varicose veins, overweight or disease of heart and arteries.

147

9.5 Posture

When standing, with contact only by the areas of the two feet, the muscles have the right degree of tension to maintain the body in the correct posture. **Fainting** results from loss of nervous control of the muscles; muscular tone ceases and the individual collapses.

Sitting also demands the correct tension of the muscles of the trunk and head. The importance of exercise for general well-being must be stressed, for incorrect posture can result in malfunction of the digestive, lymphatic and muscular systems. Continual sitting in a slumped position while reading or writing can produce back problems (see Fig. 9.7).

Lifting heavy weights (see Fig. 9.9). **Postural defects** (see Fig. 9.8).

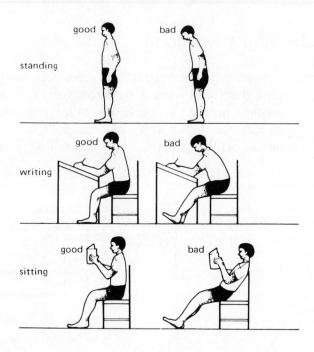

Fig. 9.7   Good and bad postures

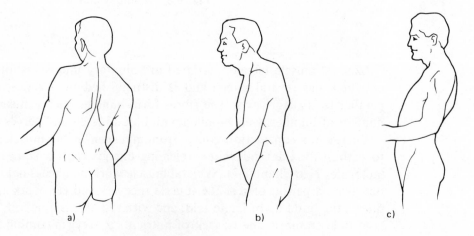

Fig. 9.8   Spinal deformities: (a) scoliosis – lateral curvature; (b) kyphosis – hunchback; (c) lordosis – hollow back

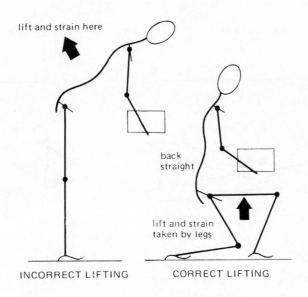

lift and strain here

back straight

lift and strain taken by legs

INCORRECT LIFTING   CORRECT LIFTING

Fig. 9.9   Lifting heavy objects

9.6   **Exercise**

1. **Muscles** should be **warmed up** with gentle exercise in order to prevent pulled muscles or torn tendons when exercise becomes more prolonged and demanding.
2. Exercise should be **regular** in order that muscles do not lose tone and preparedness. The heart and lungs become steadily stronger and are able to cope with increased demands.
3. Exercise helps to **improve the vital capacity** of the lungs (see Chapter 7). The increased air-flow enables more oxygen to be taken to the tissues and more waste gases to be returned for elimination. Appetite, rate of digestion and peristalsis are all increased by exercise.

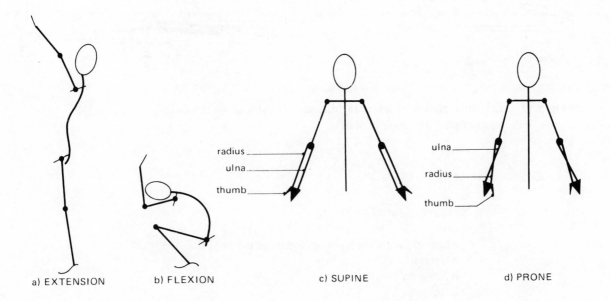

a) EXTENSION   b) FLEXION   c) SUPINE   d) PRONE

radius
ulna
thumb

ulna
radius
thumb

Fig. 9.10   Movement of bones about joints

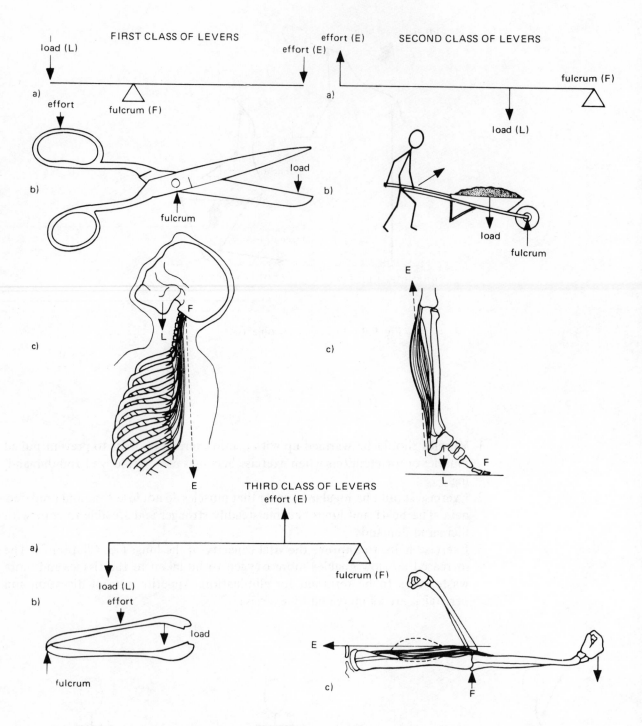

Fig. 9.11 The three classes of lever: (a) diagram; (b) simple machine using same principle; (c) example – the body of Man

9.7 Questions and answers

(a) Multiple-choice Questions

1 Which of the following bones is part of the axial skeleton?
   A femur
   B sternum
   C scapula
   D innominate
   E humerus

150

2 Which one of the following bones is part of the appendicular skeleton?
  A clavicle
  B mandible
  C coccyx
  D sacrum
  E atlas

Below is a list of some of the functions of the mammalian skeleton. Refer to these functions when answering questions **3** and **4**.

1. locomotion
2. support
3. protection
4. blood cell production
5. muscle attachment

3 Which of the following functions are performed by the humerus of a man?
  A 1, 2 and 4 only
  B 2 only
  C 3 and 4 only
  D 4 only
  E 1, 2, 4 and 5

4 Which of the following functions are performed by the vertebral column of a woman?
  A 2 only
  B 2 and 3
  C 2 and 4
  D 3 and 4
  E 2, 3 and 5

Below is a list of the different types of human vertebrae. Refer to this list when answering questions **5–7**.

1. lumbar
2. cervical
3. thoracic
4. caudal (coccyx)
5. sacral

5 To which vertebrae are the ribs attached?
  A 1
  B 2
  C 3
  D 4
  E 5

6 To which group of vertebrae do the atlas and axis belong?
  A 1
  B 2
  C 3
  D 4
  E 5

7 Of which vertebrae are there only seven in humans?
  A 1
  B 2
  C 3
  D 4
  E 5

Questions **8–10** refer to the following types of joint:
1. hinge joint
2. ball-and-socket joint
3. gliding joint
4. pivot joint

5. peg-and-socket joint
6. fixed joint

8 Which kind of joint occurs between the femur and tibia of a woman?
  A 1
  B 2
  C 3
  D 4
  E 5

9 Which kind of joint occurs between the bones of the cranium of a man?
  A 2
  B 3
  C 4
  D 5
  E 6

10 Which kind of joint occurs between the carpals of the human wrist?
  A 1
  B 3
  C 4
  D 5
  E 6

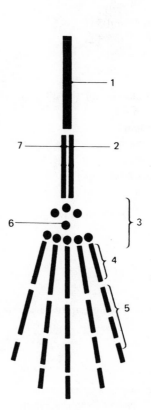

The figure above is a generalised plan drawing of a mammalian pentadactyl limb. The human limbs are modelled on this plan. Examine it carefully and then answer questions 11–14.

11 What is the name given to bone 1 in the forearm?
  A femur
  B radius
  C humerus
  D ulna
  E fibula

12 What is the name given in the hindlimb (leg) to the individual bones marked 5?
A digits
B phalanges
C metacarpals
D tarsals
E metatarsals

13 In the hindlimb (leg) of a woman, which bones are resting on the ground when she is standing normally?
A all of 3, 4 and 5
B some of 3, 4 and 5
C 4 and 5 only
D 5 only
E all of 3, 4, 5 and 6

14 If the diagram were to represent the human arm in the prone position, on which bone would the olecranon process (funny bone) be located?
A 1
B 2
C 4
D 6
E 7

15 Which one of the following is used to connect the bones of a joint in a mammal?
A ligaments
B cartilages
C tendons
D muscles
E synovial membranes

16 Which of the following bones are connected by a rotating (pivot) joint?
A humerus and scapula
B femur and tibia
C carpals of the wrist
D atlas and axis
E skull and axis

17 A bone treated with hydrochloric acid, which dissolves part of its content, can then be easily bent without breaking. Which of the following have been removed by the acid?
A ligaments
B tendons
C calcium salts
D protein fibres
E bone marrow

(b) **Structured Questions**

1 (a) The diagram below shows the upper part of the skeleton of **Man**.

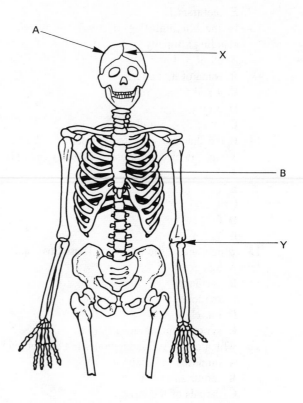

   (i) What is the name of the part labelled **A**? (one line)   **(1)**
   (ii) Name **two** structures which part **B** protects. (two lines)   **(2)**
   (iii) Which types of joint would be found at **X** and **Y**? (two lines)   **(2)**
   (iv) What is the difference in the movement of joints **X** and **Y**? (two lines)   **(2)**
(b) Vitamin D is needed for the body to allow it to absorb and use calcium from the food.
   (i) What is calcium used for in the body? (one line)   **(1)**
   (ii) If there is a shortage of foods containing vitamin D lasting several months (as often happens during wars), which part of the population would suffer most — the children, the adults or old people? (one line)   **(1)**
   (iii) Why would that group suffer most? (five lines)   **(2)**

(YHREB, 1985)

2 (a) The diagram represents a leg and its major muscles.

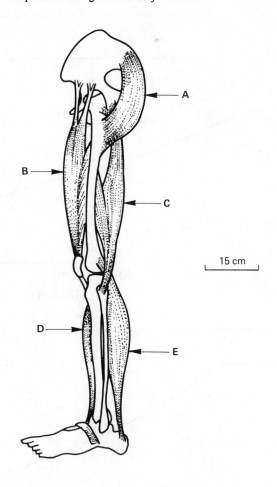

15 cm

    (i) Name the bones to which muscle **A** is attached. **(1)**
    (ii) Give **one** effect of the contraction of muscle **E**. **(1)**
    (iii) Give **two** effects of the contraction of muscle **B**. **(2)**
    (iv) Identify the muscles which are antagonistic to muscle **B**. **(1)**
 (b) (i) When the right leg is drawn back, in readiness to kick a ball on the ground, which muscles will have contracted? Explain the effect each muscle has on this movement. **(3)**
    (ii) At the same time the left leg supports the body. Explain briefly how this leg is kept rigid. **(2)**
 (c) Explain briefly how the following are moved.
    (i) Blood in the leg veins. **(4)**
    (ii) Food in the intestine. **(4)**
    (iii) Mucus in the trachea. **(2)**
(AEB, 1983)

3 (b) The effects of strenuous exercise on oxygen uptake and lactic acid concentration in the blood are shown in the graphs below.

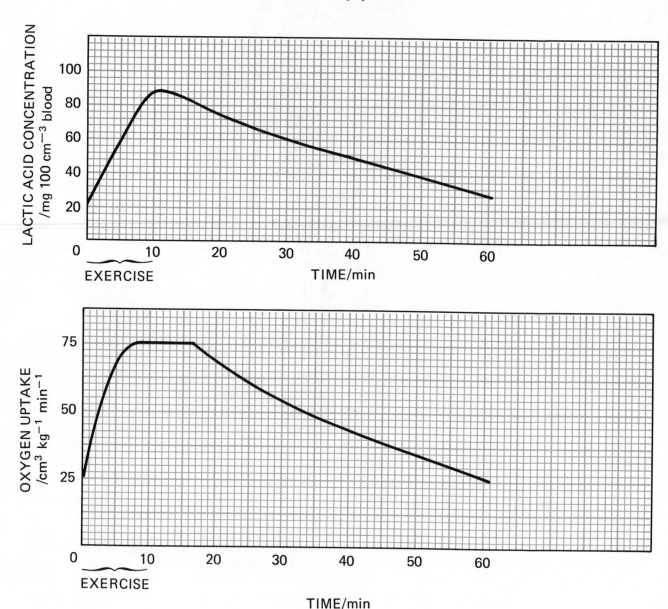

   (i) What process leads to the accumulation of lactic acid in the blood?
  (ii) How much did the lactic acid concentration increase during the period of exercise?
 (iii) Account for the level of oxygen uptake during the five minutes following the completion of the exercise.
 (iv) What substance found in muscles is involved in converting chemical energy into movement?
  (v) How is this substance restored after muscular contraction?
 (vi) Why are athletes advised to double their carbohydrate intake during the week before taking part in a marathon race?   **(9)**

(c) Illustrated below is the musculo-skeletal system of the leg of an athlete before the start of a 100 metres race.

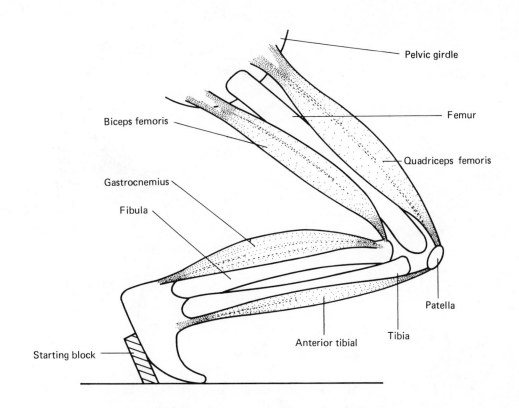

(i) Which set of muscles provides the main effort to force the athlete forward from the starting block?
(ii) Using the above diagram for reference, describe antagonistic muscle action. **(4)**

(d) (i) What is muscle tone?
(ii) What are the benefits of good muscle tone?
(iii) In what ways is the musculo-skeletal system affected by prolonged bad posture?

(5)

[Part question.] **(NISEC)**

4

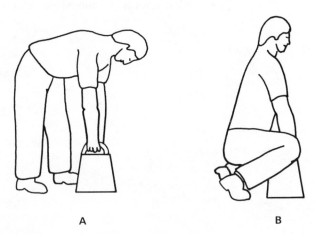

The figure above shows a section through the knee joint.

(a) Name below the parts labelled **P–T**. (5)

P (one line) Q (one line) R (one line) S (one line) T (one line)

(b) Briefly state how part **Q** differs from part **S** in its function. (three lines) (2)

(c) What is the function of the part labelled **R**? (three lines) (2)

(L)

5 The diagrams represent two ways of lifting a heavy mass.

A   B

(a) State which diagram, **A** or **B**, represents the method most likely to cause damage to the backbone.

. . . A . . . .

(b) Name the region of the vertebral column which is most likely to be damaged.

Lumbar region . . . . (2)

(AEB, 1984)

6 (a) Label the diagram of the skeleton of the arm. **(6)**

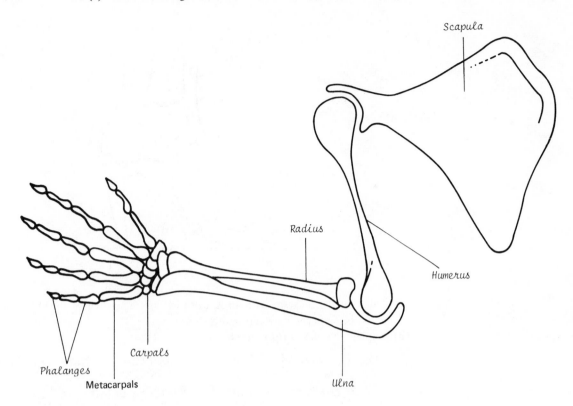

Scapula

Radius

Humerus

Phalanges

Carpals

Metacarpals

Ulna

(b) Name the muscle which, when it contracts, causes the arm to flex (bend).

Biceps muscle **(1)**

(c) Name the muscle which, when it contracts, causes the arm to straighten.

Triceps muscle **(1)**

(d) What is meant by the term 'antagonistic' when referring to a pair of muscles?

When one muscle of the pair contracts, the

other relaxes and therefore is extended. **(2)**

(e) Name the tissue found at the ends of bones and explain its function (job).

Name of tissue: *Cartilage*

Function: *Smoothness – allows friction-free movement.* **(2)**

(f) What do you understand by

(i) a slipped disc? *Intervertebral disc of cartilage is displaced.*

(ii) dislocation? *A bone is moved out of a joint.* **(2)**

(g) The elbow is an example of a hinge joint. Name **three** other types of joints in the body and give an example of where each one is found.

Joint
1. *Ball-and-socket*

Example
1. *Hip*

2. *Gliding*

2. *Wrist bones*

3. *Fixed*

3. *Sutures of the cranium*

(6)
(EMREB)

7 The simplified diagram shows a foot of an adolescent girl in a high-heeled shoe.

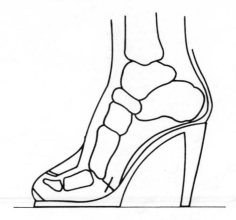

(a) Label on the diagram with an **X** to show one region where the skeleton of the foot could be damaged by wearing this shoe.

(b) Explain the reason for your choice.

The weight of the body is applied through the bones

at X instead of over the whole surface of the foot. **(2)**

(AEB, 1984)

Free-response Question

(a) List the types of joint that are to be found in the body of a mammal and give one example of each. **(5)**

(b) With the aid of a fully labelled diagram show clearly a section of a named movable joint in a mammal and its appropriate muscle attachments. **(7)**

(c) Describe how the bones of the joint are moved by the muscles and indicate on your diagram a typical nerve path by which this voluntary action is initiated and completed. **(8)**

Answer

(a)

Type of joint	Example
1. Fixed	Skull
2. Ball-and-socket	Shoulder
3. Hinge	Elbow
4. Pivot	Between atlas and axis vertebrae
5. Sliding	Wrist

(b) Section through the hinge joint of a knee (some ligaments which keep bones in place with respect to each other have been omitted)

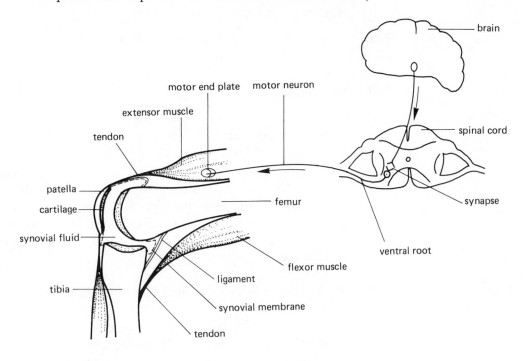

(c) The tibia will act as a lever when moved with respect to the femur. Attached to the tibia are two major muscles which together form an antagonistic pair. One of them, the flexor, can bend the limb, and the other, the extensor, straightens it. From the diagram it can be seen that, for the leg to be straightened, the extensor muscle must contract, and the flexor relax. The extensor is attached to the pelvic girdle at its uppermost end, and to the tibia/patella across a movable synovial joint at its other end.

The pelvis is fixed and immovable, and thus when the muscle shortens and thickens during contraction, the only part of the limb which can be moved is the tibia, which is pulled into line with the femur, thus straightening the limb. For the leg to revert to its former position, the reverse action has to occur; thus, the flexor contracts and the extensor relaxes and the limb returns to its bent position.

Notes

1. The question states 'list', so that time should not be wasted in writing out lengthy sentences describing the type of joint.
2. The typical nerve path has been added to the original drawing in part (b). This indicates the usefulness of reading the whole question so that you can plan your answer. Having read part (c), then space can be left to complete part (c) in part (b).

   The other part of the question can be properly answered without knowing the actual names of muscles. The terms flexor and extensor muscles have been used in their correct context to describe the movement of the joint.

   The question seems a little brief and short of words but there is a list, and there is a large drawing which can be very time-consuming.

(c) Answers to Objective and Structured Questions

(i) *Multiple-choice Questions*

1. B 2. A 3. E 4. E 5. C 6. B 7. B 8. A 9. E 10. B 11. C 12. B
13. B 14. B 15. A 16. D 17. C

(ii) *Structured Questions*

1 (a) (i) Cranium (ii) Lungs and heart (iii) **X**, fixed joint; **Y**, hinge joint
(iv) **X**, no movement; **Y**, in one plane only
  (b) (i) Formation of bone and teeth (ii) The children
(iii) Vitamin D is essential for the absorption of calcium from the gut and
thus vital for bone and teeth formation. Children are growing very
rapidly, so that their need for calcium is greatest.

2 (a) (i) Femur and pelvic girdle (ii) The foot points down.
(iii) Straightens the leg when it is flexed; raises the whole leg when it is
extended.
(iv) **A** and **C**
  (b) (i) **A** and **C** contract and pull back the femur; **E** contracts and flexes the
lower leg.
(ii) All the muscles of the left leg are under tension to lock the leg bones.
  (c) (i) By the contraction of the leg muscles against the veins the blood is
pushed upwards. Its return flow is prevented by semilunar valves in
the veins.
(ii) By peristalsis – that is, the contraction of the circular muscles of the
gut behind the food bolus. The longitudinal muscles relax. The contraction of the circular muscles moves along the gut, pushing the food
bolus in front.
(iii) By the action of a ciliated epithelium on the inner wall of the trachea.
The cilia beat in unison and carry the mucus, together with entangled
dust, upwards towards the larynx.

3 (b) (i) Anaerobic respiration (ii) 68 mg per 100 cm³
(iii) This is the oxygen debt built up during the formation of lactic acid
(anaerobic respiration).
(iv) ATP (v) ADP + phosphate + energy = ATP
(vi) To build up a store of glycogen in the muscles and liver
  (c) (i) The quadriceps and gastrocnemius
(ii) When the quadriceps contracts to straighten the leg and push off from
the block, the biceps femoralis relaxes.
  (d) (i) This is the degree of partial contraction of the muscles.
(ii) There is always a degree of tension in muscle, so that there is no danger
of damage to the muscle when a heavy load is imposed.
(iii) Prolonged bad posture will lead to extra expenditure of muscular
energy and so lead to fatigue and discomfort, e.g. lack of balance of
the head on the spine leads to strain of neck muscles.

4 (a) **P**, thigh muscle (quadriceps femoris)
Q, tendon **R**, synovial fluid **S**, ligament **T**, bone (tibia)
  (b) **Q** consists of white fibres that are inelastic.
S consists of yellow fibres that are elastic.
  (c) Cartilage at the end of the bone that, together with the same substance on
the other bone, forms a smooth articulating surface.

Questions **5–7** have the answers supplied with the question.

10 Sense Organs

10.1 Irritability

Irritability is the ability of an organism to react in response to changes in its environment. The change is called a **stimulus** and the **reaction** to it is a **response**. The stimulus is detected by **receptors** (eyes, ears, nose and tongue) and the nervous system **co-ordinates**.

10.2 The eye

The functions of the eye are as follows:

1. to judge **distances**;
2. to see the **shape** of objects in three dimensions (height, breadth and depth);
3. to **focus** clearly on objects far away and near to;
4. to see **colours**;
5. to **swivel**, to cover a field of vision.

For the external and internal structure of the eye see Figs. 10.1–10.3. The following are the functions of its parts:

1. **Sclerotic** – a tough outer coat maintaining shape.
2. **Choroid** – a pigmented black layer preventing reflection of light; through minute blood vessels it provides nutrients to the layers of the eyeball.
3. **Retina** – contains light-sensitive cells to record images.
4. **Iris** – controls the amount of light entering the eye through the pupil.
5. **Lens, ciliary muscle and suspensory ligament** – all concerned with the fine focusing of light onto the retina.
6. **Muscles** attached to the eyeball – concerned with the movement of the eyeball in the eye-socket.

In Man the two eyes are in the front of the head, separated by the nose. The result is that each eye receives slightly different pictures. The brain merges these

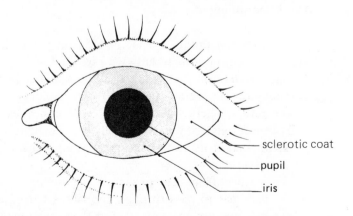

sclerotic coat

pupil

iris

Fig. 10.1 Front view of the eye of Man

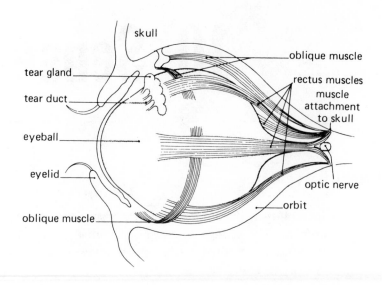

Fig. 10.2 Side view of the eyeball in the socket, showing eye muscles

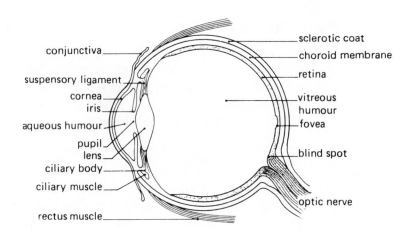

Fig. 10.3 Horizontal section through the eyeball

two pictures into one and the result is a stereoscopic picture that conveys both depth and distance.

(a) Functioning of the Eye

The eye of a mammal – that of Man, in particular – resembles a camera, except that the focusing is different (see Fig. 10.4 and Table 10.1).

The mechanism by which the eye focuses is known as **accommodation**. The accuracy of the picture received is due to light-sensitive cells called **cones**, present in the retina. There are about 3.5×10^5 cones responsible for colour vision. Other cells called **rods** are extremely sensitive to light at low light intensities. They are most important in nocturnal animals, e.g. bats. There are about 6.5×10^7 rods in each eye of Man.

Nerve fibres from the optic nerve are spread over the inner surface of the retina. Thus, light has to pass through the fibres before entering the retina. Where these fibres converge on the optic nerve there are no retinal cells, so that this point is called the **blind** spot.

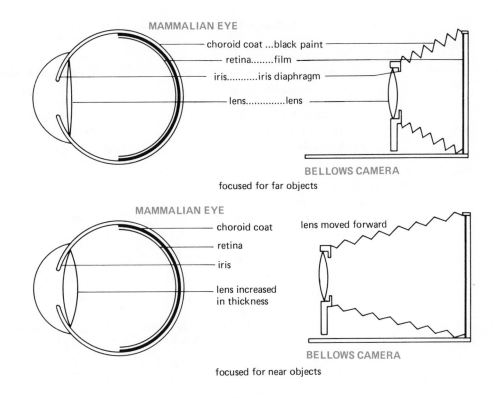

Fig. 10.4   Comparison of the eye with a camera

Table 10.1   Comparison of the eye with a camera

Mammalian eye	Camera	Function
1. Iris	Iris diaphragm	Adjusts the quantity of light entering
2. Cornea and convex lens	Convex lens	Focuses light
3. Sensitive retina	Sensitive film or plate	Detects light (image is formed here)
4. Change in thickness	Lens moves backwards and forwards	Adjusts focus for near and distant objects

The fovea or yellow spot is at the point of focus of light on the retina and contains only cones. Towards the edge of the retina there are mainly rods.

(b) Defects of the Eye

1. **Myopia** (short sight). The axis length of the eyeball may be inherently too long and as a result short-sightedness develops. Light rays are brought to a focus in front of the retina; thus, objects must be brought near to the eye in order to see them clearly. This can be corrected by **concave** spectacle lenses of the appropriate strength (see Fig. 10.5(b)).

2. **Hypermetropia** (long sight). The axis length of the eyeball may be inherently too short. Light rays are brought to a focus behind the retina and therefore, as the eye adjusts (accommodates), only distant objects are brought into

Table 10.2 Accommodation and pupil reflex of the eye to near and distant objects

Looking up from a book to a distant scene	*Looking back to the book from the distant scene*
1. Radial muscles of the iris contract; circular muscles relax Pupil gets larger	Circular muscles of the iris contract; radial muscles relax Pupil gets smaller (to concentrate light on the fovea)
2. Ciliary muscles relax Lens becomes thinner and less refractive Distant scene focused on the retina	Ciliary muscles contract Lens becomes thicker and more refractive Book print focused on the retina

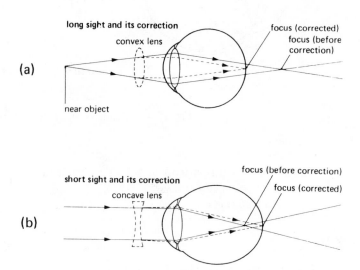

Fig. 10.5   (a) Long sight and its correction; (b) short sight and its correction

focus. This can be corrected by **convex** spectacle lenses of the appropriate strength (see Fig. 10.5(a)).

3. **Astigmatism** is the condition where the curvature of the cornea and the lens is not uniform. Thus, vertical and horizontal planes cannot be focused at the same time. The remedy is to use cylindrical lenses which refract light in one plane only and eliminate distortion.

4. **Cataract** is a condition where the eye-lens has become opaque. It occurs more commonly in old people, and the resulting blindness can be cured by removing the lens surgically. Light is still largely refracted by the cornea and this can be assisted by the appropriate spectacle lenses.

5. **Glaucoma** results from an increase in fluid pressure within the eyeball, which may damage the optic nerve and result in blindness together with considerable pain. The fluid in the eyeball normally filters out gradually, and if the exit is blocked, fluid builds up. Surgery can relieve the condition by means of a small incision in the sclerotic, allowing the fluid to filter out through the scar.

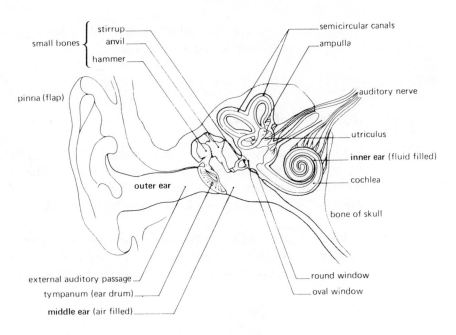

Fig. 10.6 Structure of the ear

10.3　The ear

For the structure of the ear, see Fig. 10.6.

(a)　Functioning of the Ear

1. **Detection of sound** (hearing). Sound waves, picked up and concentrated by the **pinna**, pass down the **auditory canal** and strike the **tympanum**, causing it to vibrate. These vibrations are passed across the middle ear by the ear-bones (**ossicles**), causing the **oval window** to vibrate. These vibrations are transmitted through the **cochlea** and stimulate the sensory cells. The cells generate impulses which pass through the **auditory nerve** to the brain. Low frequencies are picked up near the tip of the cochlea and high frequencies at the base. The ear can detect frequencies of 20–20 000 cycles per second (Hz) in young people, but with aging, higher frequencies are lost and the upper limit may fall as low as 5000 Hz.

2. **Detection of gravity and motion**. The inner ear has three **semicircular canals**, at right angles to one another. The **utriculus** connects the bases of the canals and it is lined with **sensory cells** connected to the nerve fibres. Entangled in the sense cells are grains of chalk (calcium carbonate). Gravity acting on these grains makes them press against the sense cells, and the resulting nerve impulses are sent to, and interpreted by, the brain. The body thus becomes **aware of its position**.

   Motion of the head is detected by the movement of fluid in the semi-circular canals. The sense cells in the **ampullae** are disturbed by the movement of the **endolymph** and send impulses to the brain. Reflex actions result and these adjustments keep the body under control during movement.

3. **The eustachian tube** connects the middle ear with the back of the throat (pharynx). This enables the air in the middle ear to be continuous with the outside air. Thus, any change in air pressure can be transmitted to the middle ear. This equalises pressure on both sides of the tympanum. Changes in air

pressure, as in, e.g., ascending or descending in an aeroplane, can be corrected to prevent damage to the eardrum.

(b) Defects of the Ear

Earache is usually caused by inflammation of the middle ear. Infection is spread by way of the eustachian tube from the throat. The infection may arise from bacteria in decaying teeth, from the after-effects of a cold or from more serious diseases, such as measles. It is important to seek medical advice for any ear infection.

Wax can accumulate in the outer ear passage, causing deafness. Again medical assistance should be sought for clearing any obstruction. On no account should pointed objects be pushed into the outer ear to clear wax.

10.4   Reception of chemical stimuli

(a) Taste

The detection of chemicals entering the buccal cavity is the function of the tongue. The chemicals dissolve in the moisture of the saliva and can then stimulate the taste buds, which are lined with sensory cells. There are only four primary tastes: **sweet, sour, salt** and **bitter.**

(b) Smell

Chemicals in the air are detected high up in the nasal cavity by sensory cells (olfactory epithelial cells). These receptor cells stimulate nerve endings, and the resultant impulses are relayed to the brain and interpreted as smell.

Man's sense of smell is very poor compared with that of some other mammals, e.g. dog and antelope. Many attempts have been made to analyse and classify smells, but so far these have not been successful.

The many **flavours** of food are the result of a **combination of taste and smell.** The ready evaporation of molecules from food, both outside and inside the buccal cavity, are detected in the lining of the nasal cavity. The blocking of the nasal channels by mucus as a result of a cold in the head effectively destroys any flavours of food.

10.5   Reception of stimuli in the skin

The skin is an important sensory organ. The receptors present are of various types, but each is connected to a nerve fibre. The following stimuli can be detected by the skin receptors: heat, cold, touch (roughness and smoothness), pain.

Different parts of the body have skin receptors much closer together in some areas than in others, e.g. the skin receptors on the fingertips are much closer together than those on the forearm. The fingertips are therefore used to feel objects and convey a great deal of information to the brain.

10.6 Questions and answers

(a) Multiple-choice Questions

1 The adjustment of the eye in a normal person in order to view either near or distant objects is called
   A refraction.
   B accommodation.
   C contraction.
   D astigmatism.
   E myopia.

2 The pupil of the eye is surrounded by
   A the retina.
   B the iris.
   C the cornea.
   D the sclerotic.
   E ligament.

3 In which of the following parts of the eye are the cones of the retina more concentrated?
   A ciliary body
   B sclerotic
   C blind spot
   D fovea
   E choroid

4 Which one of the following lenses would be effective for a short-sighted person viewing distant objects?

A   B   C   D   E

5 Older people are often seen to hold books at arm's length when they are reading. Which of the following would account for this?
   A The eyeball has become larger with age.
   B The eyeball has shrunk with age.
   C The refractive power of the lens has increased with age.
   D The refractive power of the lens has decreased with age.
   E The ciliary muscles have stretched with age.

6 Which one of the following occurs when you look up from reading a book, in shadow, in order to view a distant mountain?
   A The radial muscles of the iris contract.
   B The pupil becomes smaller.
   C The ciliary muscles contract.
   D The lens becomes thicker.
   E The suspensory ligament contracts.

7 Which one of the following comparisons between a camera and a human eye is **incorrect**?

	Camera	Eye
A	internal reflection prevented by black paint	internal reflection prevented by choroid layer
B	lens changes shape for focusing	lens moved backwards and forwards for focusing
C	light controlled by a diaphragm	light controlled by pigmented iris
D	light-sensitive chemicals in film	light-sensitive chemicals in retina
E	light enters through aperture	light enters through pupil

8 Which of the following separates the outer ear from the middle ear?
   A oval window
   B eustachian tube

C cochlea

D eardrum

E three ear bones

9 Which of the following is concerned with the passage of sound waves from the outer ear to the inner ear?

A cochlea

B eustachian tube

C semicircular canals

D pinna

E three ear bones

10 The ossicles (ear bones) are found in the

A inner ear.

B ear passage.

C cochlea.

D middle ear.

E semicircular canals.

11 Which of the following statements about the ear of Man is **not correct**?

A Changes of altitude (atmospheric pressure) affect the inner ear.

B There is a connection with the throat.

C Both ears function together to locate sound.

D It can distinguish between vibrations of different frequencies.

E It is an organ of balance.

12 When an astronaut is outside the gravitational field of the earth, there is a disruption of the normal functions of the

A ear ossicles and tympanum.

B cochlea and endolymph.

C semicircular canals and ampullae.

D oval and round windows.

E cochlea and ear bones.

13 The eustachian tube.

A equalises pressure between the oval and round windows.

B prevents the ear ossicles from amplifying vibrations.

C equalises pressure on either side of the tympanum.

D transmits vibrations to the cochlea.

E magnifies the vibrations of the tympanum at the oval window.

14 Which one of the following is the reason for the strong smell of substances that evaporate readily into the air?

A They stimulate the taste buds and inhibit the olfactory nerve.

B They stimulate the nerve fibres to the brain.

C They prevent the taste buds from working.

D They have an anaesthetising effect on the brain.

E They produce small particles which stimulate the sense cells and the olfactory nerve.

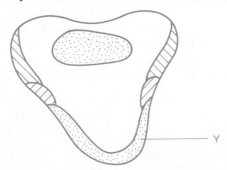

15 In the above figure which one of the following is the taste sensation of position **Y**?

A saltiness

B bitterness

C sweetness

D sourness

E none of the above

16 Receptors in the ampullae of the semicircular canals are stimulated when

A wax is present in the ear.

B the oval window vibrates.

C pressure is equalised in the middle ear.

D the round window vibrates.

E the head sharply rotates.

17 Which of the following may account for partial deafness in the ear when there is an accumulation of wax?

A Air pressure is not balanced by the eustachian tube.

B The inner ear fluids are put under extra pressure.

C The oval window becomes rigid.

D The tympanum (eardrum) is unable to vibrate properly.

E The middle ear ossicles cannot move, owing to wax.

(b) Structured Questions

1 Read through the following account of hearing in Man and write in the appropriate word or words on the dotted lines to complete the meaning of each sentence.

In Man, sound waves are funnelled into the outer ear by the . . . . These sound waves cause the . . . . to vibrate and the vibrations are carried across the cavity of the . . . . by a chain of . . . . The vibrations reach a second membrane called the . . . . which vibrates against the fluid contained in the spiral-shaped . . . . Since fluids cannot be compressed the inward bulge of this membrane is compensated by the movement of the . . . . into the middle ear. In the inner ear vibrations passing through the endolymph stimulate . . . ., and nerve impulses developed from them are sent to the brain along the . . . .

2 (a)

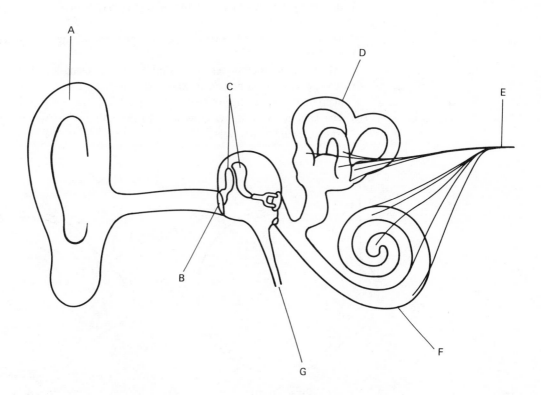

Which of the letters in the above diagram of a human ear, labels:

    (i) a structure that converts sound waves to mechanical vibrations? (one line) **(1)**

    (ii) a structure sensitive to changes in the position of the head? (one line) **(1)**

    (iii) a structure which converts vibrations to nerve impulses? (one line) **(1)**

(iv) a structure which allows the air pressure in the ear to equalise with atmospheric pressure? (one line) **(1)**

(v) a structure for transmitting nerve impulses to the brain? (one line) **(1)**

(YHREB, 1984)

3 (a) The graph shows the average hearing ability for persons of different ages. For the purposes of this graph the hearing ability of a ten-year-old child is taken as 100%.

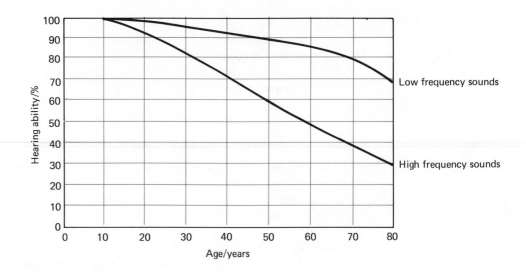

(i) What does this graph tell you about the change in hearing ability as a person becomes older? (three lines) **(3)**

(ii) What is the percentage hearing ability for high frequency sounds of a 60-year-old person compared with a ten-year-old child? (one line) **(2)**

(b) Briefly explain why each of the following may reduce the hearing ability of a person of any age:

(i) a large mass of wax forming at the inner end of the external ear canal. (two lines) **(2)**

(ii) the three small bones of the middle ear are fusing together. (two lines) **(2)**

(c) It is possible for a person to know the direction from which a sound is coming. Explain how the presence of two ears on the head makes this possible. **(1)**

(d) The simplified diagrams show the eye in section when viewing a distant object and when viewing a near object.

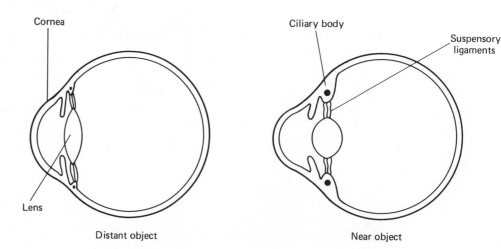

Distant object   Near object

Give four ways in which the eye viewing the distant object is different from the eye viewing the near object, as shown by the diagrams. (4 × two lines) **(4)**

(e) Finish the table to show one different function for each of the parts of the eye named.

Part of eye	Function
Iris	
Choroid	
Cone cells	
Yellow spot (fovea)	

(f) The simplified diagram shows an eye, in section, with the rays of light uncorrected for the condition known as myopia (short sight).

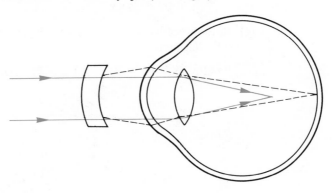

On this diagram draw in the path of the rays of light after they have passed through the correcting lens until they reach the back of the eye. **(2)**

(SREB)

4 (a) The diagram below shows a simple camera. The human eye is like a camera in many ways. Look at the diagram and then complete the chart with the corresponding parts of the eye. (The first one has been done for you.)

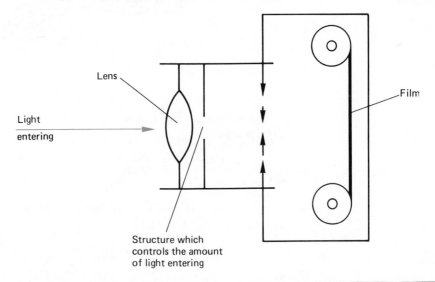

Camera	Eye
1. Lens of glass	Lens made of cells
2. The hole through which light enters	*Pupil*
3. Structure which controls the amount of light entering	*Iris*
4. The film at the back of the camera	*Retina*

(3) 173

(b) A short-sighted man can read the words of a book clearly **only** when it is held at point A from the front of the eyeball, **B**. The line below represents this distance.

A ├──────────────────────────────────┤ B

(i) What is the actual length of the line **AB**? (in cm) .... *6.2 cm* ... **(1)**

(ii) If every cm on the line **AB** represents 4 cm, at what distance must the book be held for clear vision? (Show your working.)

6.2 x 4 = 24.8 cm
.... **(2)**

(iii) What type of correcting lens would be needed if the book was held 200 cm away?

Convex lens
.... **(1)**

(c) In the space below, complete the diagram of the **front** view of the eye and label the following parts.

EYEBROW, EYELID, SCLEROTIC COAT, IRIS, PUPIL

— *eyebrow*

— *eyelid*

— *sclerotic coat*

— *pupil*

— *iris*

(8)

(d) Write down the function (jobs) of each of the following.

(i) Eyebrows.

Protection - channels away rainwater to side of head.

(ii) Eyelashes.

Protective and sensory

(iii) Eyelids.

To shut out light and distribute tear fluid.

(iv) Tears.

Discharge of tear gland to keep surface of eyeball moist.

(v) Retina.

Inner layer of eyeball sensitive to light. .... **(5)**

(EMREB)

5 The human skin contains touch receptors. The table below shows the results of an experiment in which ten pupils each measured the minimum distance between two sharp points at which they could be felt separately on different areas of the skin.

Pupil number	1	2	3	4	5	6	7	8	9	10
Minimum distance (mm) two points felt on skin of:										
(a) Fingertips	2	3	4	1	2	2	2	3	4	2
(b) Palm of hand	11	9	8	10	9	9	11	13	11	9
(c) Lips	1	2	1	1	1	2	1	1	1	1
(d) Forearm	17	10	15	16	14	16	15	18	10	9

(a) (i) Calculate the average minimum distance between the points, at which they both can be felt, for each area of skin in the ten pupils.

Fingertips .... *2.5* .... .mm

Palm of hand .... *10.0* .... mm

Lips .... *1.2* .... mm

Forearm .... *14.0* .... mm **(4)**

(ii) Explain why it is better to take an average of the results from ten pupils rather than use the result from a single person.

Because individuals may vary and a single person could give a distorted result. Furthermore, in scientific experiments it is important to obtain as much data as possible. **(2)**

(iii) State **two** conclusions about touch sensitivity which can be made from a study of these results.

1. *The lips are the most sensitive of the four areas.*

2. *The forearm is the least sensitive.* **(2)**

(b) Name **two** stimuli, other than touch, which can be detected by receptors in the human skin.

(i) *Temperature*

(ii) *Texture* **(2)**

(c) The retina of the human eye contains light receptors. Describe by what means
(i) light from an object is correctly focused on the retina.

Light is refracted by the cornea such that it is converged onto the lens. The final, fine focusing is brought by the lens changing shape. **(2)**

(ii) Impulses (messages) pass from the light receptors to the brain.

> *Impulses pass by way of the optic nerve. Owing to the action of sodium ions, the impulse passes as a change in voltage.* **(1)**

(UCLES)

6 (a) The human eye can undergo changes to focus on both near and distant objects. What is the name of this process?

> *Accommodation* .... **(1)**

(b)

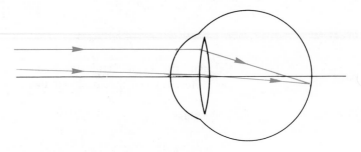

The diagram above shows an eye focused on a distant object. Draw a similar diagram in the space below to show an eye focused on a near object.

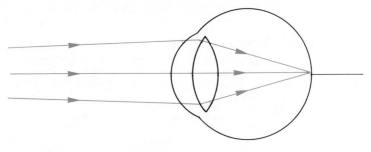

(4)

(c) Explain briefly how light is prevented from being reflected inside the eye.

> *The middle layer of the eyeball, the choroid, is black and thus prevents reflection.* **(2)**

(d) Account for the fact that an object cannot be seen if its image falls upon the blind spot.

> *The sense cells, rods and cones, have nerve endings on the inner surface of the retina. These nerve fibres pass over the surface of the retina to the optic nerve, which pierces the retina (no rods or cones)* **(3)**

(L)

Free-response Question

(a) Draw a fully labelled diagram of the ear. **(8)**

(b) Describe how the ear is adapted for hearing. **(8)**

(c) What do the eye and the olfactory organs of the nose have in common with the ear, in terms of structure and function? **(4)**

Answer

(a)

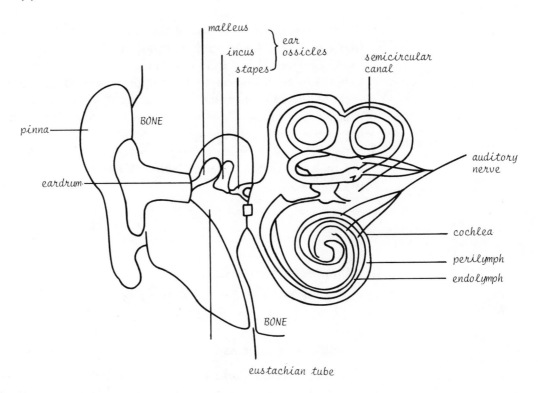

(b) The pinna helps to collect and channel sound waves into the outer ear. It can also assist in judging the direction from which the sound has come, although this function is limited in man compared with other mammals. Sound waves which enter the ear set the eardrum vibrating and the vibrations are conveyed through the three bones of the air-filled middle ear on to the oval window. The eardrum has a greater surface area than the oval window, and this together with the lever action of the ear bones (ossicles) magnifies the force of the vibrations by a factor of about twenty-two.

As the stapes hits the oval window, it sets the fluid of the inner ear in motion, especially in the cochlea. These oscillations pass down the cochlea, and in this area there are sensory cells which respond to the fluid movements. Vibrations of the highest frequency are detected in the first part of the cochlea and those of the lowest frequency in the last part. The range from high to low is continuous and gradual from one end of the cochlea to the other.

When stimulated by a particular frequency, impulses are passed to the brain via the auditory nerve and interpreted there as a particular sound.

The middle ear is filled with air and is usually at a pressure equal to that of the atmosphere. If changes in air pressure take place outside the eardrum, the pressure can be adjusted by the eustachian tube opening and admitting more air, or releasing excess air from the middle ear, as the case may be. Thus, pressure on either side of the eardrum becomes equalised and in this way possible rupture of the delicate eardrum is prevented. Such rupture or distortion could lead to deafness.

(c) All three sensory structures possess sensory cells which are able to respond to specific stimuli. They are all capable of translating their respective stimuli into nerve impulses. These impulses generated in the sensory tissues are conducted to the brain by sensory nerves, and the brain is thus able to detect environmental changes. This will enable the body to respond in the appropriate manner.

Notes

1. The drawing is probably one of the most difficult to produce under examination conditions. With practice, however, this drawing can be learned and full marks obtained if it is fully labelled.
2. Part (b) of the question simply requires a description of the function of the parts of the ear concerned with hearing. Each structure must be described as it passes the sound waves of the air from structure to structure, finally starting a nerve impulse in the auditory nerve.
3. Note that part (c) must be very short, since the previous two parts take up the majority of the time available. This is indicated by the mark allocation (4).

(c) Answers to Objective and Structured Questions

(i) *Multiple-choice Questions*

1. B   2. B   3. D   4. A   5. D   6. B   7. B   8. D   9. E   10. D   11. A   12. C
13. C   14. E   15. A   16. E   17. D

(ii) *Structured Questions*

1 Words to be inserted in order of appearance below:
ear trumpet/pinna; eardrum/tympanum; middle ear; ear ossicles/ear bones; oval window; cochlea; round window; sensory cells; auditory nerve
2 (a) (i) B   (ii) D   (iii) F   (iv) G   (v) E
3 (a) (i) Percentage of low-frequency sounds decreases by 30%.
          Percentage of high-frequency sounds decreases by 70%.
      (ii) 50%, compared with 100%
   (b) (i) The wax prevents the correct vibration of the eardrum (tympanum) and thus the transmission of sound through the ear ossicle of the middle ear.
       (ii) The movement of the ear bones transmits and amplifies sound from the eardrum to the oval window. If the bones are fusing, this transmission cannot occur.
   (c) The ears receive the sound from slightly different angles and the resolution of this difference within the brain can indicate the position of the origin of the sound.
   (d) In the eye viewing a distant object: the lens is thinner; the cornea is less curved; the ciliary body is smaller (relaxed ciliary muscles); the suspensory ligaments are taut.
   (e) Iris, to control the size of the pupil
       Choroid, to supply nutrient to the eye
       Cone cells, to detect different wavelengths of light as colour
       Yellow spot (fovea), most cones are present giving clearest vision
   (f) See dotted lines on the figure in question 3(f).

Questions **4–6** have the answers supplied with the questions.

11 Co-ordination

11.1 Introduction

The sense organs (see Chapter 10) receive stimuli which must be co-ordinated by the **nervous system** and the **endocrine system** (see Table 11.4). The impulses produced in the nervous system travel in some nerves at 65 metres per second or 140 miles per hour. The endocrine system discharges hormones into the blood system, and these produce effects in target organs or tissues at a much slower rate.

(a) Organisation of the Nervous System

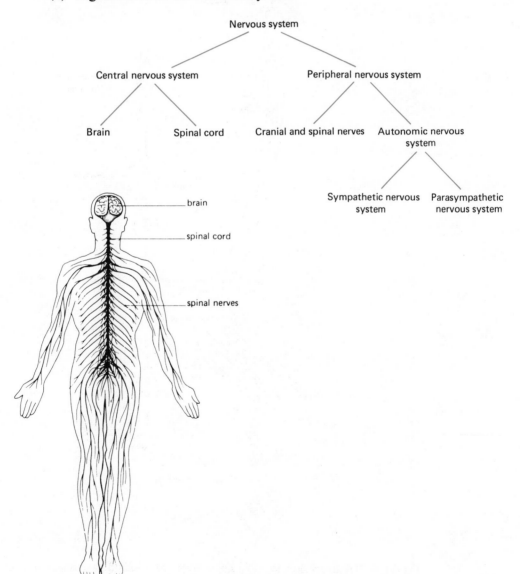

Fig. 11.1   Diagram of the nervous system of Man

11.2   Nerves

For the structure of neurons and nerves and their functions see Figs. 11.2 and 11.4.

An **impulse**, produced by minute electrical charges, travels along the **axon** of the neuron. It travels in one direction, and does so by **depolarisation** of the axon over a small area which causes the inside to become positive and the outside negative. Depolarisation lasts only for one-thousandth of a second at any one point, but as resting polarisation develops again, the next point becomes depolarised. Thus, the impulse travels as a **wave of depolarisation**. The impulse reaches the end of a fibre, and then there is secreted minute amounts of a chemical, **acetylcholine**. This chemical moves across the **gap (synapse)**, to produce a new impulse in the adjoining neuron.

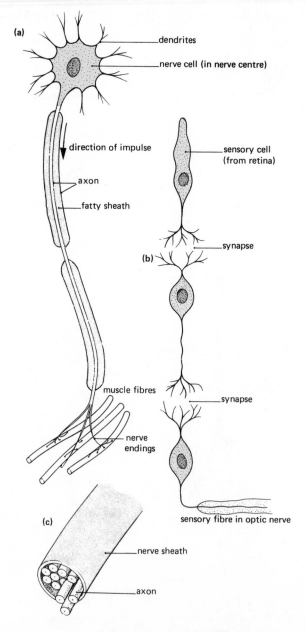

Fig. 11.2   Diagrams showing (a) a single motor neuron and its connection with muscle fibres; (b) synapses between a sensory neuron and a sense cell of the retina of the eye; (c) fibres of neurons bound together to form a nerve (*after D. G. Mackean*)

Sensory neurons conduct impulses from **sense organs** to the **central nervous system**. **Motor neurons** conduct impulses from the central nervous system to **effector organs** such as muscles or glands. Notice the different positions of the cell bodies in Fig. 11.2. **Connector** (intermediate or internuncial) neurons carry impulses up and down the spinal cord. They are located in the white matter of the brain and spinal cord.

11.3 The brain

The main regions of the brain are as follows.

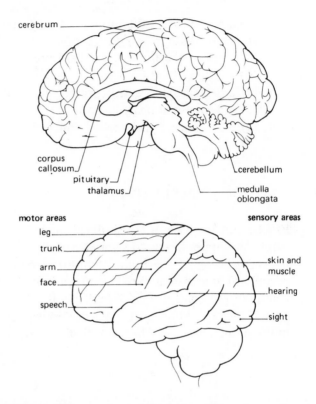

Fig. 11.3   Longitudinal median section of the brain of Man. Areas of control in the cerebrum of the brain of Man

1. The **olfactory lobes** are connected by sensory neurons to the organ of smell, which also sends neurons to the cerebrum.
2. The **cerebrum** (cerebral hemispheres) has a much-folded,wrinkled appearance and forms the bulk of the brain. The outer cortex is formed of grey matter (cell bodies of neurons); the inner part is white matter (axons of neurons). Different sensory areas of the cerebral cortex control sight, hearing, smell and skin sensation, while motor areas control muscles of the legs, arms, face, eyes and head. Other areas, missing in all other animals, are concerned with Man's intellectual functions, including speech, music, mathematics and abstract thought.
3. The **hypothalamus** provides reflex control concerned with homeostatic mechanisms such as temperature control, water control and carbon dioxide levels in the blood.
4. The **thalamus** relays information from lower centres of the brain to the cortex.

5. The **optic lobes** receive sensory neurons from the eyes; they are small in Man, since the cerebral cortex has taken over much of their role.

6. The **cerebellum** is concerned with the maintenance of balance, location and positioning of the body, i.e. co-ordination of muscular activity so that all movements are performed smoothly.

7. The **medulla oblongata** is the reflex centre of the brain controlling blood pressure, coughing, swallowing, sneezing, yawning and vomiting. Inside the centre is thought to be the region concerned with '**wakefulness**'. The activity of the medulla oblongata's neurons enables the cerebral cortex to control conscious activities such as seeing, moving, talking, etc. When the 'wakefulness' centre **becomes inactive** we fall **asleep** and then the following physiological changes take place:

(a) a reduction in the number of impulses in the cerebral cortex;
(b) an increase in sweat secretion;
(c) a fall of 1°C in body temperature;
(d) a fall in blood pressure;
(e) a decrease in breathing rate;
(f) a decrease in heart beat;
(g) an increase in the rate of conversion of glucose to glycogen in the liver.

11.4 Injuries to the central nervous system

Fracture of the base of the skull can cause fatal brain damage. Injuries to the cerebrum are more likely to cause paralysis. Concussion results from a blow and, hence, bruising of the brain. Symptoms range from a headache and giddiness to periods of unconsciousness up to a month in length. Damage to the spinal cord results in the condition of paraplegia in which there is paralysis and loss of sensation below the injury, e.g. in legs, bowels and bladder. In quadriplegia all four limbs are paralysed; it can result from injuries to the upper spinal cord.

11.5 Influence of drugs on the central nervous system

1. **Stimulants** increase the activity of the CNS, e.g. amphetamine (pep pills), caffeine in tea or coffee, benzedrine and nicotine in tobacco.
2. **Depressants** decrease the activity of the CNS, e.g. alcohol, aspirin, tranquillisers, barbiturates, opium and its derivatives (morphine and heroin).

See Section 15.2.

11.6 Reflex action

A reflex action is an **automatic**, unlearned **response** to a **stimulus**. The true reflexes of Man are very limited and occur soon after birth, e.g. suckling and clinging to the thumb of an adult. There are, however, certain protective reflexes in the adult, such as (1) blinking to protect the eye, (2) sneezing to clear the nasal passages, (3) coughing to remove phlegm. For the **differences** between a **spinal reflex** and a **voluntary action** see Table 11.1. For the **path** of a spinal reflex action, see Fig. 11.4. For the **pathway and speed** of a reflex action, see Fig. 11.5.

Table 11.1 Comparison of a spinal reflex action and voluntary action

Spinal reflex	Voluntary action
Stimulus affects external or internal receptor	Initiated from the brain at the conscious level
Spinal cord only involved – not under the control of the will	Forebrain involved – under the control of the will
The impulse travels only up or down the spinal cord	The impulse travels from the brain down the spinal cord
The path of the nerve impulse is by the shortest route	The path of the nerve impulse is much longer
The response is immediate	The response can be delayed
The response is in skeletal or internal involuntary muscle	The response is in skeletal muscle only

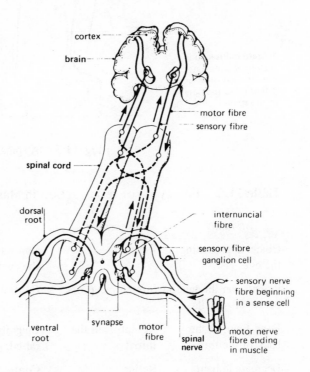

Fig. 11.4 Diagram of a reflex arc, together with a portion of the spinal cord and the nerve connections to the brain

A reflex action can be summarised as follows:

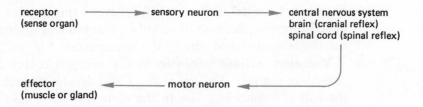

A **cranial reflex** involves the brain and the cranial nerves, e.g. blinking, action of the iris to control light in the eye, sneezing, etc.

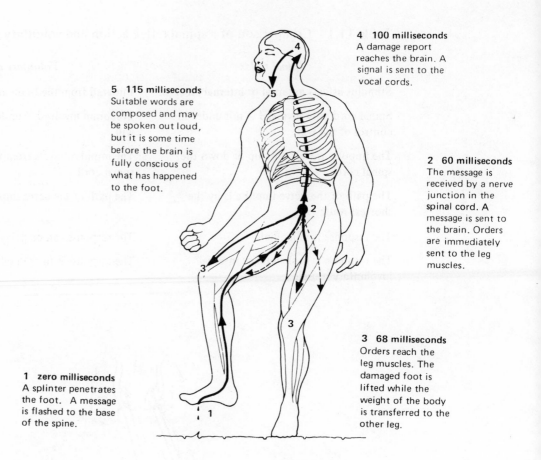

Fig. 11.5 A spinal reflex

1 zero milliseconds
A splinter penetrates the foot. A message is flashed to the base of the spine.

2 60 milliseconds
The message is received by a nerve junction in the spinal cord. A message is sent to the brain. Orders are immediately sent to the leg muscles.

3 68 milliseconds
Orders reach the leg muscles. The damaged foot is lifted while the weight of the body is transferred to the other leg.

4 100 milliseconds
A damage report reaches the brain. A signal is sent to the vocal cords.

5 115 milliseconds
Suitable words are composed and may be spoken out loud, but it is some time before the brain is fully conscious of what has happened to the foot.

Table 11.2 Five types of reflex action in Man

Stimulus	*Receptor*	*Response*	*Reflex action*
Object approaching the eye	Retina	Contraction of muscles of the eyelid	Blinking
Pressure on the tendon below the patella	Stretch receptors	Contraction of flexor muscles	Knee-jerk
Food bolus in the back of the throat	Receptors in the throat	Epiglottis closes; soft palate raised; peristalsis	Swallowing
Decrease in light intensity	Retina	Contraction of radial muscles of the iris	Pupil dilates
Increase in light intensity	Retina	Contraction of circular muscles of the iris	Pupil constricts

A **spinal reflex** involves the spinal cord and the spinal nerves, e.g. dropping a hot plate, knee-jerk reflex, etc. Whenever a spinal reflex occurs, we are still aware of it happening, because of impulses that are sent up the spinal cord to the brain, but the brain does not control the response (see Fig. 11.5).

Voluntary actions originate in the cerebral cortex of the brain and involve conscious thought. For example, a boy decides to kick a football. The eye sees the ball and sends impulses to the cortex. In the cortex a decision is made, based on past experience. The cortex initiates impulses down the spinal cord, spinal nerves and nerve endings such that the correct muscles are stimulated to contract in the leg. The foot then kicks the ball.

11.7 The conditioned reflex

The conditioned reflex was first demonstrated by a Russian physiologist, **Ivan Pavlov**, experimenting with dogs. He fed meat to his dogs and measured the amount of saliva produced – a normal reflex action. He then rang a bell every time meat was produced and conditioned the dogs to expect meat whenever they heard the bell. After many trials the dogs **would salivate in response to the ringing of the bell only**. Thus, a conditioned reflex is an **example of a response** (salivation) **produced by different nerve pathways from those of the original stimulus**.

The conditioned reflex can form the basis of training by reward and punishment. Young children are conditioned in a similar fashion by their parents, sometimes by reward (sweets), sometimes by punishment (a slap). Many complex activities, such as walking, swimming or riding a bicycle, are performed without thinking consciously of each stage of our actions. That is, they become similar to conditioned reflexes but, in addition, there is intelligent learning involved and they cannot be considered as simple, isolated, conditioned reflexes.

11.8 The autonomic nervous system

The autonomic nervous system is a collection of nerves and ganglia controlling the internal activities of which the individual is not normally aware, i.e. unconscious activities. These include peristalsis, glandular activity and heart beat. Two sets of nerves – sympathetic and parasympathetic – which connect the CNS to the internal organs (see Table 11.3).

Table 11.3 Comparison of sympathetic and parasympathetic nervous systems

Sympathetic	*Parasympathetic*
Causes:	Causes:
increased rate of heart beat	decreased rate of heart beat
decreased rate of gut peristalsis	increased rate of gut peristalsis
constriction of arteries	dilation of arteries
constriction of anal sphincter	dilation of anal sphincter
constriction of bladder sphincter	dilation of bladder sphincter
relaxation of bladder wall	contraction of bladder wall
dilation of bronchioles	constriction of bronchioles
dilation of pupil	constriction of pupil
transmitter substance is noradrenaline	transmitter substance is acetylcholine
ganglia alongside spinal cord	ganglia without effector organs

Table 11.4 Comparison of nervous and endocrine activity

	Nervous control	*Endocrine control*
Stimulus	Through receptors, eyes, nose or internal receptors, include light, gravity, sound, temperature, etc.	Through external or internal receptors
Linking mechanism	Central nervous system and nerves	Blood and circulatory system
Effectors	Muscles and glands	Whole body, organs or organ systems
Speed	Rapid reaction – reflex arc or voluntary nerve paths	Slow for some such as growth; rapid for others such as fight or flight hormone, adrenaline

11.9 The endocrine system

The endocrine system works in parallel with the nervous system. The differences between the two systems are shown in Table 11.4. Endocrine or ductless glands pass their products directly into the bloodstream. Each hormone acts upon a particular target tissue or organ (see Table 11.5).

Table 11.5 Endocrine glands and their functions

Gland	Position in body	Hormone secreted	Response of body to hormone	Abnormal functions
Thyroid	Neck	Thyroxine	Controls basic metabolism and growth rate	Deficiency causes dwarfism and mental retardation in childhood, myxoedema in adult. Overproduction causes increased metabolism
Islets of Langerhans	Pancreas (dual function – exocrine and endocrine)	Insulin	Controls balance of sugar in the blood	Deficiency results in diabetes mellitus
Adrenal gland	Attached to kidneys	Adrenaline (medulla)	Controls response for 'fight or flight', i.e increased heart beat, increased blood sugar; dilates coronary artery, pupils, etc.	
		Cortisone and other hormones (cortex)	Release glucose from protein in stress. Control salt balance in the body	
Ovary	Dorsal abdominal wall	Oestrogen	Controls growth of uterus, hip skeleton, underarm hair, pubic hair, breasts, (secondary sexual characteristics)	Deficiency causes delay of appearance of these changes
(See Section 12.2)				
Testis	In the scrotum	Testosterone	Controls growth of hair on the pubis, under arms, and on chest and face, increased muscular develment, deepening of voice	Deficiency causes delay or lack of development of these changes
Intestinal wall	Duodenum	Secretin	Controls secretion of digestive juices from the pancreas (exocrine function of this gland). Produced when acid food enters the intestine	
Pituitary	Beneath the brain	Several hormones	Controls activity of other ductless glands	Deficiency causes a type of dwarfism, and inactivity of certain endocrine glands

11.10 Questions and answers

(a) Multiple-choice Questions

1 Which of the following is the function of the fatty sheath?
   A to conduct impulses
   B to transmit information across synapses
   C to insulate the axons
   D to protect the cell bodies
   E to contact surrounding cells

2 Which of the following is the speed at which an impulse travels along a nerve fibre?
   A 12 m s^{-1}
   B 120 m s^{-1}
   C 2000 m s^{-1}
   D 200 m s^{-1}
   E $12\,000 \text{ m s}^{-1}$

3 Which of the following statements is **not** correct regarding a neuron?
   A The myelin sheath contains a fat.
   B The impulses are conducted electrically.
   C The neuron can vary considerably in length.
   D The axon and sheath form the nerve.
   E It can synapse with other neurons.

4 Impulses are transmitted across synapses by which of the following?
   A electrical means
   B mechanical means
   C chemical means
   D nuclear means
   E thermal means

Questions **5–11** refer to the following regions of the brain:
   A optic lobes
   B cerebrum
   C hypothalamus
   D cerebellum
   E medulla oblongata

5 Which one of the above is particularly well developed in mammals and contains the region concerned with speech in Man?

6 Which one of the above is a reflex centre concerned with temperature regulation?

7 Which one of the above is concerned with balance, locomotion and body position?

8 Which one of the above is concerned with memory?

9 Which one of the above is the reflex action centre of the brain controlling sneezing, coughing and swallowing?

10 Which one of the above receives sensory neurons from the eyes but whose controlling role has been largely taken over by another region of the brain?

11 Which one of the above is the 'wakefulness' centre of the brain, in that its neuronal activity enables the cerebrum to control conscious activities?

Questions **12–19** refer to the following hormones:
   A thyroxine
   B adrenaline
   C oestrogen
   D secretin
   E insulin

12 Which one of the above controls the secretion of digestive enzymes by the pancreas?

13 Which one of the above is produced in the ovary?

14 Which one of the above is produced by glands located immediately anterior to the kidneys?

15 The condition known as cretinism, in which mental and physical development is retarded, occurs in children who have an insufficient production of one of the above hormones. Which one?

16 Which one of the above contains iodine?

17 Which one of the above converts glycogen to glucose?

18 Which one of the above controls sugar balance in the blood?

19 Which one of the above is produced as a result of conditions of stress?

20 Endocrine glands produce hormones to co-ordinate activity in the body and distribute them by which one of the following?
  A tubular ducts
  B nerve fibres
  C bone cells
  D blood system
  E alimentary system

21 Which one of the following glands increases in size until puberty and then gradually diminishes?
  A thyroid
  B pituitary
  C adrenal
  D thymus
  E pancreas

22 Which of the following endocrine glands is associated with both structure and function of the brain?
  A thyroid
  B adrenal
  C thymus
  D parathyroid
  E pituitary

23 Which one of the following is the path of a spinal reflex action?
  A receptor–motor neuron–relay neuron–sensory neuron–effector
  B receptor–relay neuron–sensory neuron–motor neuron–effector
  C receptor–sensory neuron–motor neuron–relay neuron–effector
  D sensory neuron–receptor–relay neuron–motor neuron–effector
  E receptor–sensory neuron–relay neuron–motor neuron–effector

24 Which one of the following is an effect of the release of adrenaline into the bloodstream?
  A Speeds up the rate of growth of the skeleton.
  B Stimulates the growth of hair in certain regions of the body.
  C Causes regrowth of the lining of the uterus.
  D Increases the flow of blood to the skeletal muscles.
  E Causes the change of glucose to glycogen.

25 Which one of the following is a hormone produced by the pancreas.
  A testerosterone
  B insulin
  C oestrogen
  D thyroxine
  E adrenaline

26 Which one of the following glands produces digestive enzymes as well as a hormone?
  A pancreas
  B thyroid
  C sweat
  D thymus
  E liver

27 Which one of the following is not involved in the reflex action of the body known as the knee-jerk reflex?

 A spinal cord
 B motor neuron
 C sensory neuron
 D brain
 E muscles

28 Which one of the following is a human reflex action?

 A eating
 B running
 C talking
 D blinking
 E singing

(b) Structured Questions

1 The diagram represents a section through the spinal cord showing a part of a reflex arc.

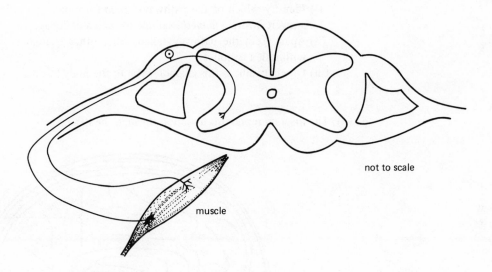

not to scale

muscle

On the diagram

(a) complete the motor neuron,
(b) by means of an arrow show the direction of an impulse along it and
(c) label a synapse.

(3)
(AEB, 1983)

2 (a) The simplified drawings below show two different side views of the human brain.

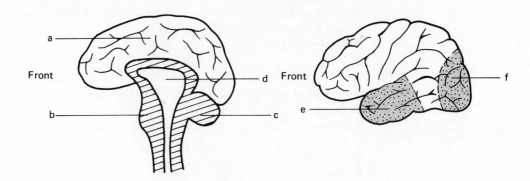

   (i) Name the structures or regions labelled **a**, **b**, **e** and **f**.
   (ii) State **one** function for each of the regions labelled **c** and **e**.
   (iii) What is found in region **d**?   **(5)**
(b) The four sequences below show possible pathways, at least one of which could form part of a reflex arc.
   Pathway **A**: motor neuron → brain → sensory neuron.
   Pathway **B**: sensory neuron → spinal cord → motor neuron.
   Pathway **C**: sensory neuron → spinal cord → receptor.
   Pathway **D**: cone cell → brain → eyelid.
   (i) Identify which of the pathways could form part of a reflex arc, supporting your answer(s) with a *named* example of such a pathway.
   (ii) Apart from the pathway taken, what other characteristic is usually associated with reflex actions?
   (iii) Of what value is this characteristic to the body?   **(6)**
                                                                                                            (NISEC)

3 (a) The diagram represents a section through the head.

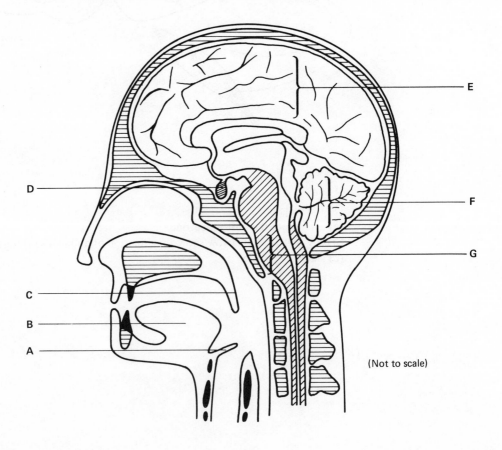

(Not to scale)

(i) Identify the structures labelled **A, B** and **C** **(3)**

(ii) Explain the role of each of the structures, **A, B** and **C**, during swallowing. **(7)**

(iii) **A** person in an upside down position can pass food along the oesophagus against the pull of gravity. Explain how this is possible. **(6)**

(b) (i) Identify gland **D** and name the **two** hormones it produces which directly affect the ovaries. **(3)**

(ii) Briefly explain the role of these hormones in the menstrual cycle. **(4)**

(c) Identify the parts labelled **E, F** and **G** and for each briefly describe its major function. **(7)**

(AEB, 1985)

4

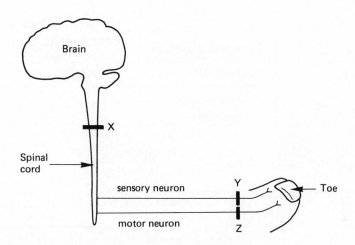

The diagram above shows, in a simplified way, the nerve supply to a toe. Lines marked **X, Y** and **Z** show regions of the nervous system which might be blocked.

(a) Complete the table below to indicate which block would produce the condition described.

In the box provided write

X – if the block is at **X**
Y – if the block is at **Y**
Z – if the block is at **Z**
N – if there is no block

Condition	Position of block
(i) The toe can be moved, but the movement cannot be felt.	Y
(ii) A pin prick to the toe can be felt, but the toe cannot be moved.	Z
(iii) When the toe is stimulated, the stimulus can be felt and the toe can be moved; the person knows it is moving.	N
(iv) When the toe is stimulated, it moves, but this movement cannot be felt.	X
(v) When the skin of the toe is stimulated, this stimulus can be felt but the toe cannot be moved.	Z

(5)

(b) A nerve impulse passes from one neuron to another across a gap.

(i) What is this gap called?

Synapse .... **(1)**

(ii) How is a nerve impulse transmitted across this gap?

The impulse arrives at the synapse and a chemical, acetylcholine, is formed. This moves across the gap and initiates another impulse in the dendrites of another neuron. **(3)**

(L)

5 Complete the table to compare some features of the nervous and hormonal control systems of the body.

	Nervous coordination	*Hormonal coordination*	
Form of transmission	*Electrical and chemical*	*Chemical*	(2)
Route of transmission	*Neurons*	*Blood system*	(2)
Relative speed of transmission	*Fast*	*Slow*	(2)

(AEB)

6

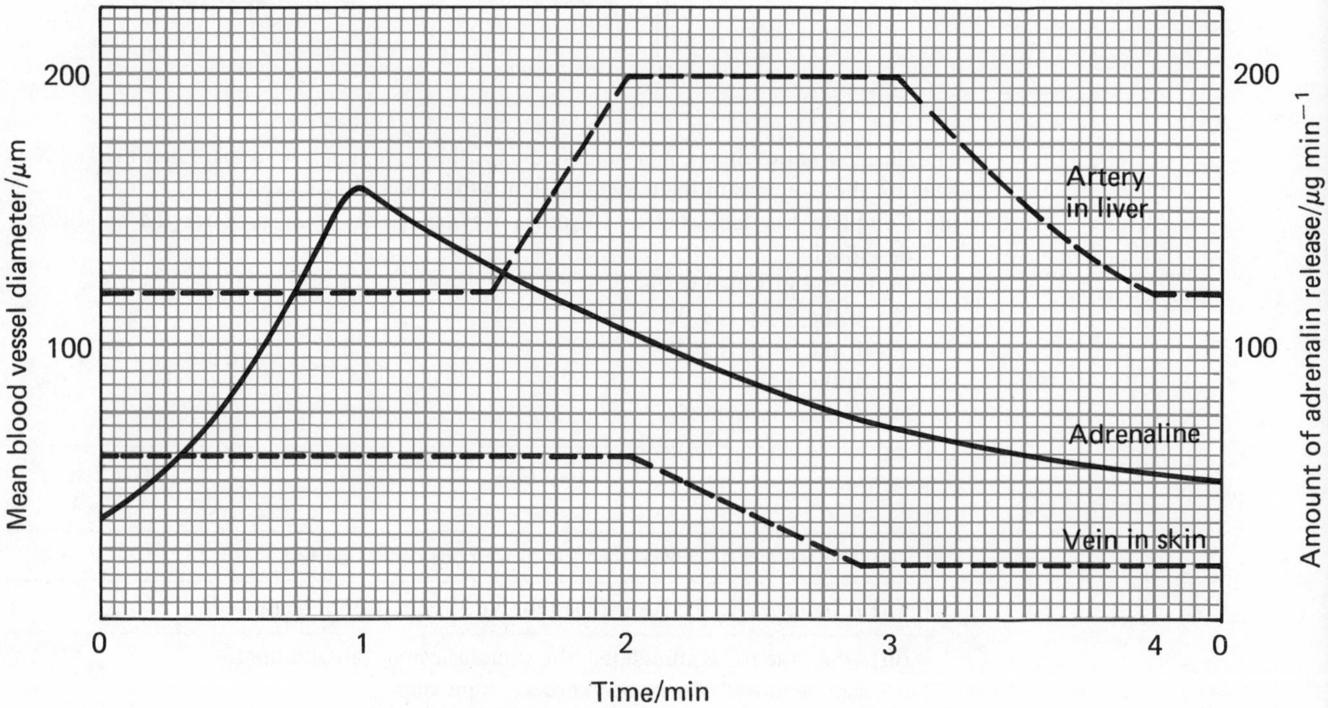

Time/min

The graph shows both the amount of adrenaline released into the bloodstream and the changes in diameter of two sets of blood vessels over a short period of time.

(a) (i) What is the maximum change in the diameter of the liver blood vessel?

200 - 120 μm = 80 μm **(1)**

(ii) What general effect would this have on the flow of blood through this region?

Increase the rate of blood flow (almost double it). **(1)**

(b) What is the time delay between the maximum production of adrenaline and the maximum effect on the vein in the skin? *2.9 - 1 = 1.9 minutes* **(1)**

(c) State **two** differences between the effects of the adrenaline on the two blood vessels which are shown on the graph.

1. *Artery increases; vein decreases in diameter.* ....

2. *Artery increase twice that of vein (80 μm: 40 μm).* .. **(2)**

(d) Suggest **two** events which could have happened at time zero resulting in the increased release of adrenaline.

1. *Lining up for a 100 metre sprint.* ....

2. *Entering an examination hall.* .... **(2)**

(NISEC)

7 The figure shows some of the structures involved in a simple reflex action. The skin of the hand is about to be pricked by a sharp drawing-pin and this will result in the forearm being moved upwards in the direction shown by the arrow.

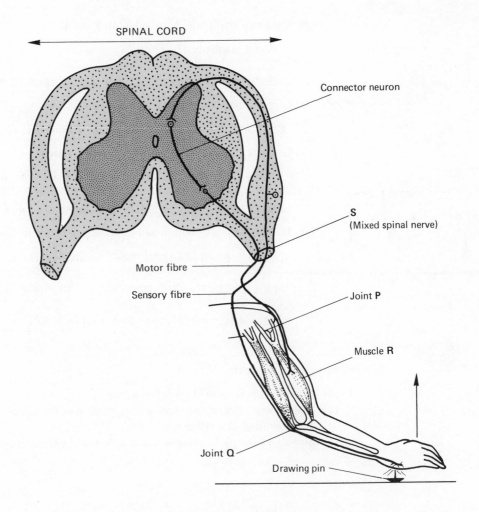

(a) (i) What must muscle **R** do in order to bring about the upward movement of the forearm?

Contract .... **(1)**

193

(ii) Describe how the impulse (message) from pain receptors receiving the stimulus is relayed to muscle **R**.

The impulses initiated by pain receptors pass along sensory neurons to the grey matter of the spinal cord. They are passed to a connector neuron and thence to motor neurons. The motor neurons convey the impulses down the spinal nerve and they reach muscle R. **(4)**

(b) Suggest a reason why structure **S** is called a **mixed** spinal nerve.

It contains both sensory and motor neurons. **(1)**

(i) State **two** differences between joint **P** and joint **Q**.

1 *P is a ball-and-socket joint; Q is a hinge joint.*

2 *P movement in several planes; Q movement in one plane.* **(2)**

(ii) State **two** similarities between joints **P** and **Q**.

1 *Both synovial joints.*

2 *Both allow movement by tendons and muscles.* **(2)**

(d) (i) Give another example of a simple reflex action.

Closure of the pupil when exposed to bright light. **(1)**

(ii) Give **one** example of a **conditioned** reflex action.

A dog (or man) salivating at the sound of a bell instead of the sight (or smell) of food. **(1)**

(iii) In what way is a conditioned reflex action different from a simple reflex action?

In the conditioned reflex action the stimulus has been changed but produces the same response. The nerve pathways are different. **(2)**

(UCLES)

Free-response Questions

1 (a) Explain what is meant by a reflex action. **(2)**
(b) Make a large diagram to show a cross-section of the spinal cord and the arrangement of the neurons in a reflex arc. **(12)**
(c) Give two examples of simple reflex actions and explain how they can be useful to us. **(6)**

Answer

(a) A reflex action is an automatic response to a stimulus.
(b) This diagram should be similar to the front part of Fig. 11.4, which shows a cross-section of the spinal cord and its accompanying sensory and motor neurons, together with a diagrammatic representation of the sense and motor cells.

(c) (i) When an object such as a hand or insect approaches the eye, it blinks. This involves the eyelid moving down rapidly in an attempt to prevent damage to the eyeball.

(ii) If one attempts to pick up a very hot object, the sense cells in the skin detect the high temperature. In a reflex response the muscles of the arm and hand either drop the object or withdraw the hand from the object. This relaxation and contraction of certain muscles is a reflex action designed to protect the skin from burning.

Note that part (c) asks for only two examples of reflex actions. There are others that could be stated besides the two mentioned above. They are stated briefly below.

(iii) The iris reflex; reducing light intensity (closing the pupil) or increasing light intensity (opening the pupil), thus protecting the retina or helping it to see better.

(iv) Rapid inhaling of air, and then coughing, to remove phlegm.

(v) Rapid exhaling of air, by sneezing, to clear the nasal passages.

Notes

1. Section (b): this is a quite complicated diagram and therefore should be kept simple and large with clear indication of areas of tissue (e.g. grey matter and white matter) and structures (e.g. the spinal nerve with its neurons). It must be drawn with speed. There are numerous labels and they will probably carry 8/9 of the 12 marks. The remaining 3/4 marks will be for the correctness of the diagram.

2. There are a number of alternatives for section (c), so that there should be no problem in remembering two, each of which gains three marks. These are clearly designated in (iii) to (v).

2 (a) Name the two systems which effect the co-ordination of the body. What similarities and differences are there between their modes of action? **(6)**

(b) Name two ductless glands, and give an account of the effects which the secretions of each have on the body. **(7 + 7)**

Answer

(a) The two systems which effect the co-ordination of the body are the nervous and the endocrine systems. Both of these systems receive stimuli by way of either external or internal receptors. The mechanism which connects the receptors to the effectors in the nervous system is the central nervous system and peripheral nerves, whereas the blood circulatory system connects the equivalent structures of the endocrine system.

The effectors of the nervous system are muscles and glands, whereas in the endocrine system the effector may be the whole body, organ systems or simply organs.

Response to endocrine stimulation can be slow and prolonged (e.g. thyroxine, promoting growth) or rapid (e.g. adrenaline). However, response to nervous stimulation is normally rapid, either by a reflex arc or by a voluntary nerve path. The nature of the nerve impulse is electrochemical, whereas the hormones of the endocrine system are purely chemical.

(b) Two ductless glands are (i) pancreas (islets of Langerhans); (ii) adrenals.

(i) The pancreas possesses endocrine cells which secrete insulin. It controls the use of sugar in the body by regulating how much is converted to glycogen and how much is oxidised for energy liberation. If the glucose (blood sugar) level in the blood is high, this stimulates the pancreas to produce more insulin. However, if the glucose level is low, less insulin is secreted.

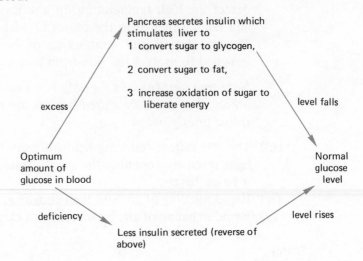

Blood sugar regulation by insulin

Hence, the level of blood sugar itself controls the amount of insulin secreted into the blood. This is an example of homeostasis.

If insufficient insulin is produced by the body, then the individual is unable to convert blood sugar to glycogen, and much of it is excreted in the urine. The disease diabetes will occur, which can be fatal if untreated. Correction of the disease is by regular injection of insulin. If oversecretion of insulin occurs, too much sugar is converted to glycogen and this may induce a coma.

(ii) The adrenal glands consist of an outer area, the cortex, and an inner one, the medulla. The cortex secretes several hormones, one being cortisone. This accelerates the conversion of protein to glucose.

The medulla secretes adrenaline. This region is connected to the nervous system, and when impulses which are associated with a situation involving stress, and needing a vigorous response, are transmitted to the brain, motor impulses are relayed to the medulla, which releases adrenalin into the blood. Adrenaline increases the heart beat and breathing rates, raises the blood sugar level, dilates the pupils, directs blood where it is most needed, and tones up the muscles. These changes therefore increase the organism's efficiency in a stressful situation. Any deficiency in secretion of adrenaline will result in an organism which is slow to react in emergencies. If there is an oversecretion, then the organism will over-react under stress.

Notes

1. Section (a) asks for a comparison of the nervous and endocrine systems. It is an unusual request, and if the student has not thought of this before he may have difficulty. Differences are clear enough but similarities need some consideration, because they are not immediately obvious. Think of the broad function of each system in this case.

2. Section (b) is a simple recall question depending on detailed knowledge of two ductless glands and their products. The syllabus usually asks for a limited number, so that this should not be too difficult.

(i) *Multiple-choice Questions*

1. C   2. B   3. D   4. C   5. B   6. C   7. D   8. B   9. E   10. A   11. E   12. D
13. C   14. B   15. A   16. A   17. B   18. E   19. B   20. D   21. D   22. E
23. E   24. D   25. B   26. A   27. D   28. D

(ii) *Structured Questions*

1 (a), (b) and (c)

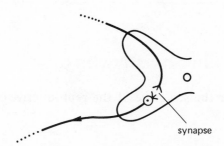

synapse

2 (a) (i) a, cerebrum; b, medulla; e, frontal lobe; f, posterior lobe   (ii) c, muscular co-ordination; e, hearing   (iii) Fluid
   (b) (i) Pathway **B** – picking up a hot plate
       (ii) Automatic response to a specific stimulus
       (iii) Protective/avoids injury
3 (a) (i) **A**, epiglottis; **B**, tongue; **C**, soft palate
       (ii) The tongue moves food around during cutting and chewing by the teeth. At the same time the food is moistened and formed into a ball. This is passed to the back of the throat. The soft palate closes the entrance to the nasal passage. The presence of the food ball causes an automatic swallowing reaction. The epiglottis closes the opening of the glottis into the larynx, thus preventing food passing into the larynx and trachea.
       (iii) When the food ball enters the oesophagus, it is passed along by a process of peristalsis. This involves the rhythmic contraction of the involuntary muscles of the gut wall. The circular muscles contract behind the food (longitudinal muscles relax) and force the food along. This wave of contraction continues upwards as muscles contract and relax.
    (b) (i) **D**, pituitary; FSH, follicle stimulating hormone, and LH, luteinising hormone
       (ii) FSH stimulates the growth of ovarian follicles. The ovum is released from the follicle (ovulation). LH stimulates the production of the corpus luteum. Following ovulation FSH and LH secretion decreases. If fertilisation does not occur, menstruation takes place and the whole cycle begins again.
    (c) E, cerebrum; F, cerebellum; G, medulla
       The cerebrum is concerned with learning, memory, reasoning, conscience and personality. The cerebellum is concerned with co-ordination of muscular movement. The medulla is concerned with automatic adjustment of body functions such as heart rate, breathing, etc.

Questions **4–7** have the answers supplied with the questions.

12 Reproduction and Growth

12.1 The reproductive structures

For the structure of the reproductive organs and gametes see Figs. 12.1, 12.2 and 12.3.

(a) Male

The **scrotum** contains the **testes**, which produce sperms and male hormone, **testosterone**. The testes are at a temperature 2°C cooler than the rest of the body. The **epididymis** is a long coiled tube alongside each testis — for sperm storage. The **sperm ducts** conduct sperm from the epididymis. The **prostate gland, seminal vesicles** and **Cowper's gland** are found at the junction of sperm ducts and urethra. The glands secrete fluids (nutrients and enzymes) onto the sperms and the resulting mixture is called semen. The **penis** transmits sperms to the female through the urethra (a duct with urinogenital functions — see Chapter 8). Prior to sexual intercourse, the penis dilates, hardens and becomes erect, owing to spongy tissue filling with blood.

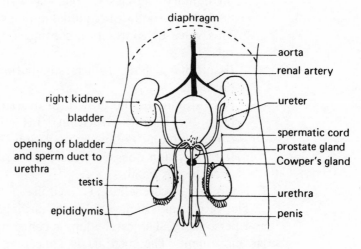

male reproductive organs (position in body)

Fig. 12.1 Male reproductive organs

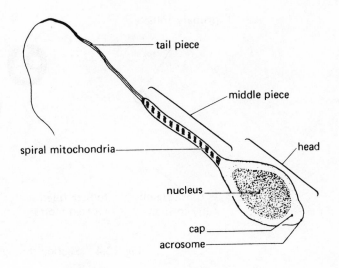

Fig. 12.2 Drawing of a sperm cell

(b) Female

The **ovaries** produce **ova** (eggs) and female hormones (**oestrogens** and **progesterone**). The **oviducts** conduct the ova down to the uterus. The single **uterus** (womb) is muscular and small (80 mm x 10 mm), but can expand considerably during pregnancy. The **cervix** is a ring of muscle at the entrance to the womb from the vagina. The **vagina** is a muscular tube opening into the **vulva**, enclosed by lips (**labia**). At the ventral end of the vulva is a small structure, the **clitoris**. The vagina and the clitoris are well supplied with sensory cells. The **urethra**, also opening into the vulva, carries urine from the bladder; it takes no part in sexual activity (unlike the urethra of the male).

(i) *Ovulation*

See Figs. 12.4 and 12.5.

One ovum per month is produced from alternate ovaries. The ovum is wafted into the oviduct by ciliated cells, where it is still surrounded by small cells called the **corona radiata** (see Fig. 12.5). The cells of the **Graafian follicle**, having released the ovum, continue to grow, filling the space of the follicle. The new tissue is

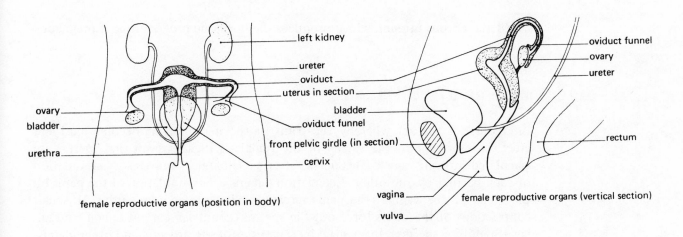

Fig. 12.3 The reproductive system of a female human

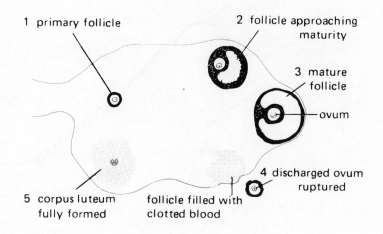

Fig. 12.4 Section through an ovary

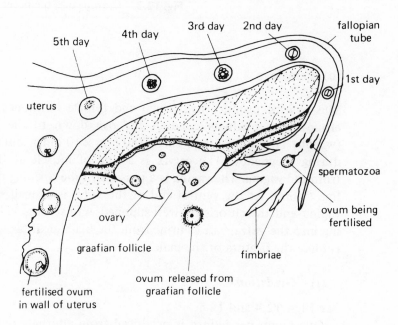

Fig. 12.5 Ovulation, fertilisation and implantation. Note that the lining of the womb is thickened and well supplied with blood. Had fertilisation not occurred, the lining would have broken down and the blood flowed away during menstruation

called the **corpus luteum**, which produces the hormone progesterone if pregnancy occurs.

12.2 Intercourse and fertilisation

Intercourse is the act whereby the penis of the male, having become hard and erect, is inserted into the vagina of the female, with the result that sperms are ejaculated into the vagina. This occurs after psychological and sensory stimulation of each partner by the other. Ejaculation occurs after stimulation of the penis by the walls of the vagina (the penis moving backwards and forwards). Muscular contractions of the ducts force out the sperms (about 1.5 cm^3 of semen containing 100 million sperms). Influenced by secretions of the prostate and other glands, the sperms swim through the cervix into the uterus and up each oviduct. If they reach an ovum (at the correct time of the month — see Section 12.6), one sperm penetrates the egg membrane. Other sperms are prevented from entry and the two nuclei (from sperm and egg) fuse, forming a zygote. Thus, fertilisation has taken place and this can occur at any time in the 24 hours after copulation (see Fig. 12.5).

12.3 Pregnancy

For a summary of development of the foetus see Table 12.1.

Table 12.1 Stages in the development of a human baby

Time after fertilisation	Length	Stage of development
7–10 days	140–160 µm (diameter)	Hollow ball of cells, thickened in one area and implanted in the uterine wall
3 weeks	1.5 mm	Head region obvious. Spinal cord and heart starting to develop
6 weeks	10 mm	Brain growing rapidly. Eyes and ears developing. Arm and leg buds forming
12 weeks	9 cm	The embryo (now called a foetus) has almost the external appearance of a miniature baby
9 months	50 cm	Birth (parturition) occurs (see Fig. 12.8)

The fertilised egg (**zygote**) undergoes division in the oviduct to form a **ball of cells**. At the same time it is moved down the duct, by beating cilia, into the uterus, and becomes embedded (**implantation**) in the wall of the uterus. A series of minute projections (**villi**) grow between the embryo and the uterine wall. These gradually enlarge and form the **placenta**, within which the maternal and embryonic blood systems will be closely intertwined (see Fig. 12.7).

The placenta is partly maternal in the uterine wall and partly embyronic tissue, and is connected to the embryo by the umbilical cord. It has a large internal surface area, with the maternal blood vessels forming pools, or **sinuses**. The membranes of these sinuses are very thin and are adjacent to the embryonic capillaries. These features aid rapid transfer of substances:

glucose, oxygen, amino acids, salts, vitamins pass **from mother to embryo**;
urea, carbon dioxide and other waste products pass **from embryo to mother**.

The maternal and the embryonic blood systems always remain separate, for the higher blood pressure of the maternal circulation would burst the delicate blood vessels of the embryo (foetus).

As the placenta develops, further membranes completely surround the embyro. These include the **amnion**, containing the **amniotic fluid**. Thus, the embryo and later the foetus live in a fluid for nine months until birth. The fluid **protects** the foetus from shock and **allows it to move freely** during growth.

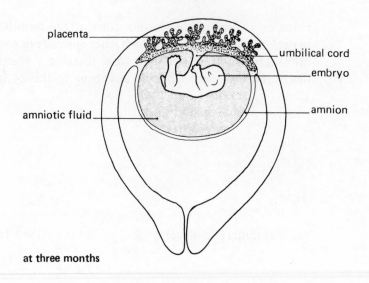

at three months

Fig. 12.6   An early foetus in the uterus (10 weeks)

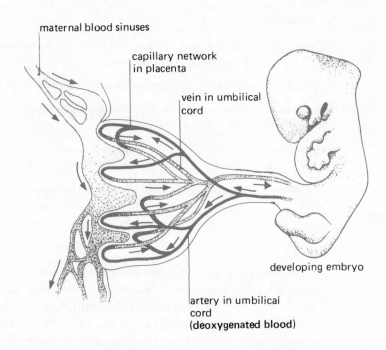

Fig. 12.7   Foetal and maternal circulation in the placenta

(a) The Placenta as a Barrier

1. Bacteria cannot pass across the placenta but some **viruses** do so, e.g. rubella (German measles) virus. This affects the nervous system of the foetus, which may be born with brain damage as well as being blind or deaf.
2. **Alcohol** can pass across the placenta to the foetus. The mother should avoid consumption of alcohol. The child may be born underweight, or even with alcohol poisoning, if the mother is an alcoholic.
3. **Nicotine and other chemicals** inhaled during tobacco smoking can pass to the foetus and lead to reduced birth weight.
4. **Drugs** can pass into the bloodstream of the foetus from the mother. Mothers must ensure that any drugs prescribed during pregnancy are incapable of

affecting the foetus. The 'hard' drugs heroin and cocaine, if used by the mother, can have grave effects on the foetus.

5. **Rhesus factor** (see Chapter 6).

12.4 Birth

1. The foetus turns **head downwards**.
2. The **muscles of the uterus contract**, the contractions becoming rhythmic and more and more frequent; the cervix opens.
3. The **amnion breaks**, releasing fluid (breaking of waters).
4. **Uterine contractions** become more powerful; the head emerges from the cervix, then the rest of the foetus; the foetus passes through the vagina.

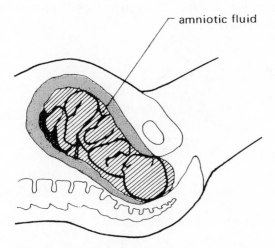

amniotic fluid

1 The amnion is about to rupture ("breaking of the waters")

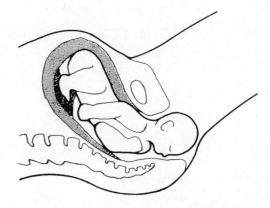

2 The uterine wall contracts, forcing out the head

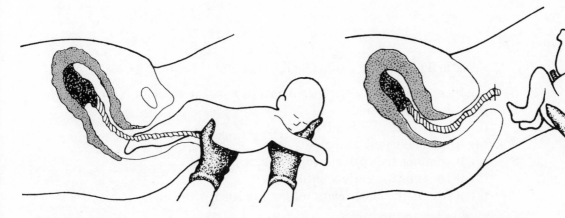

3 The baby is born

4 The umbilical cord is tied and cut

Fig. 12.8   Stages of birth

5. At this stage the baby must **pass head first** (the largest diameter of the body) through the **pelvic girdle** (see Chapter 9).
6. The baby emerges from the vulva and begins to breathe air.
7. The umbilical arteries and veins are cut in the umbilical cord; circulation now becomes adult (see Chapter 6).
8. The placenta separates from the uterine wall, passes through the uterus and the vagina, and is expelled as the afterbirth.

(a) Early Stages of Care by Mother

1. **Milk** is produced from the **mammary glands** (breasts). **Colostrum**, produced for the first 2–3 days, is a thin liquid containing proteins and antibodies. The latter protect the gut of the baby against infections. The colostrum is followed by milk (see Table 12.2 for composition). Note the much higher protein content in cow's milk, for the faster growth of the calf.

Table 12.2   A comparison of cow's milk and human milk

	Cow's milk (per kg)	Human milk (per kg)
Water	860 g	876 g
Protein	34 g	12 g
Fat	42 g	36 g
Carbohydrate	48 g	70 g
Vitamin C	15 mg	52 mg
Vitamin A	0.4 mg	4.5 mg
Vitamin D	0.0002 mg	0.0003 mg
Sodium	0.6 g	0.15 g
Potassium	1.4 g	0.6 g
Magnesium	1.3 g	0.35 g
Calcium	1.2 g	0.3 g
Phosphorus	1.0 g	0.15 g
Energy	2730 kJ	2730 kJ
pH	6.8	7.3

Natural feeding by mother's milk is desirable because:

(a) It contains the correct ingredients for a balanced diet (except iron).
(b) It is easily digested.
(c) No sterilisation is needed.
(d) It is immediately available.
(e) It contains the correct proportion of protein.
(f) It contains the correct antibodies.
(g) It fosters the close relationship of mother with child.

Artificial feeding with cow's milk does not have the above advantages, but must be used where the mother cannot feed the child from her own milk.

2. **Weaning** is the withdrawal of breast or bottled milk and its replacement with prepared semi-solid food. It normally begins at about four months.

3. At six months **harder foods**, such as biscuits, rusks, soft fruit and cheese, can be given when gums are hard and chewing is possible. Teething rings at this stage help the hardening of gums and the chewing process.
4. A **balanced diet** should be provided, based on the same principles as for an adult (see Chapter 4). Milk properly sterilised still provides the basis of the diet.
5. **Sleep requirements** gradually reduce from 24 hours a day to about 14 hours a day at one year of age. Morning and afternoon rests are gradually decreased until at 4–5 years the need is for about 12 hours per night. This continues up to 10–11 years.

All young babies should be adequately covered with night clothes and bed clothes, to compensate for heat loss at night (see Chapter 8). For the first few years a warm bedroom is desirable during winter months, to prevent **hypothermia** (see Chapter 8).

12.5 Growth

Growth is the increase in size of an organism or its parts due to synthesis of protoplasm or external cellular substances (e.g. matrix of bone and cartilage, fibres of tendons, etc.).

Boys and girls show a period of rapid growth in the first two years, then a steady increase until **puberty**, when a further increase in growth is shown (see Fig. 12.9). The growth **rate** shows a remarkable increase at puberty. At puberty the change from child to adult becomes apparent, with the development of secondary sexual characteristics (Table 12.3).

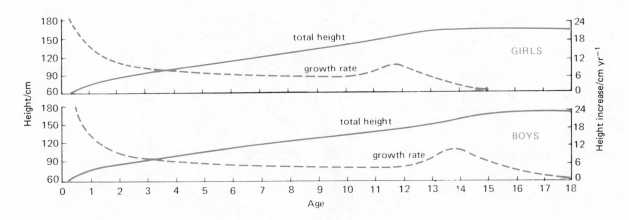

Fig. 12.9   Human growth rates

Table 12.3   Secondary sexual characteristics

Girl	*Boy*
1. Rapid growth; height and weight	1. Rapid growth; height and weight
2. Breasts, uterus grow; hips widen	2. Muscles and penis grow
3. Hair grows; armpits and pubic region	3. Hair grows; armpits, pubic region, face and chest
4. Menstruation begins	4. Sperms produced
5. Secretion of oestrogen and progesterone	5. Voice lowers in pitch ('breaks')
6. Emotional changes; attachment to opposite sex	6. Emotional changes; attachment to opposite sex

12.6 Menstruation

For hormonal control of the menstrual cycle, see structured question 6. For a graphical representation of the menstrual cycle, see the figure in structured question 6.

The purpose of the menstrual cycle is to **prepare the lining of the uterus** to become swollen with blood, ready for the implantation of a fertilised egg. If fertilisation does not occur, the thickened lining breaks down and the blood flows out through the vagina. This is known as the **menstrual period**, and lasts about five days. During this time the blood is absorbed by cotton **sanitary towels** or pads, which should be changed often. Fully grown girls and women may use **tampons**, which are cylinders of absorbent material that fit inside the vagina.

12.7 Questions and answers

(a) Multiple-choice Questions

1 A fertilised ovum is known as
   A a gamete.
   B a zygote.
   C an egg.
   D an oocyte.
   E an oosphere.

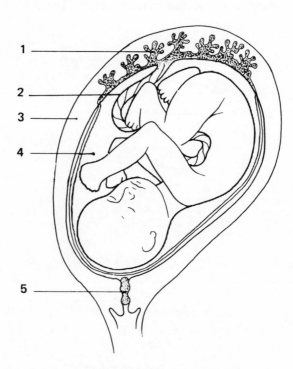

Study the above figure and answer questions **2–6**.

2 Approximately what stage in foetal development is shown?
   A 1 month
   B 3 months
   C 4 months
   D 5 months
   E 8 months

3 Where is the amniotic fluid located?

   A 1

   B 2

   C 3

   D 4

   E 5

4 What does 5 represent?

   A mucus

   B blood vessels

   C semen

   D placenta

   E chorion

5 What happens to the structure labelled 1 after parturition?

   A It remains in the uterus and is absorbed.

   B It becomes detached from baby and uterus and remains floating.

   C It is passed out of the vagina, leaving the uterine wall intact.

   D It is passed out of the vagina, together with part of the uterine wall.

   E It remains in position and is used by later foetuses for nutrition.

6 Which of the following is the most complete description of the contents of the structure labelled 2?

   A an artery

   B a vein

   C an artery and a vein

   D an artery, a vein and a lymph vessel

   E an artery, a vein and a tube connecting the stomachs of mother and foetus

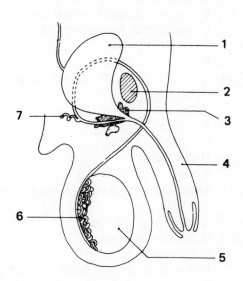

Study the above figure and answer questions **7–11**

7 Which one of the following structures contains bone?

   A 2

   B 3

   C 4

   D 5

   E 6

8 Where are the spermatozoa produced?

   A 1

   B 3

   C 5

   D 6

   E 7

9 Where are the spermatozoa stored?

A 2
B 3
C 5
D 6
E 7

10 Where is the urine stored?

A 1
B 2
C 4
D 5
E 7

11 Which one of the following does **not** contribute to the formation of semen?

A 3
B 4
C 5
D 6
E 7

12 Which one of the following attaches the foetus to the mother's uterus?

A placenta
B umbilical cord
C navel
D spinal cord
E foetal membranes

13 The number of chromosomes in each ovum is

A the same as in a spermatozoon.
B one more than in a spermatozoon.
C one less than in a spermatozoon.
D twenty-three pairs.
E twenty-four pairs.   (UCLES)

14 The flexibility of the skeleton of an unborn baby is due to the fact that it is made mainly of

A ligament.
B muscle.
C tendon.
D cartilage.
E bone.

15 Twins which are born as a result of the fertilisation of two eggs by two sperms are certain to

A have a single afterbirth.
B be of the same sex.
C be identical twins.
D develop in common foetal membranes.
E have separate placentae.

16 Which one of the following changes occurs only in boys at puberty

A growth of hair in the armpits
B widening of the pelvic girdle
C growth of breasts
D enlargement of the larynx
E growth of pubic hair

208

1 (a) The diagram shows an embryo in the uterus shortly before birth.

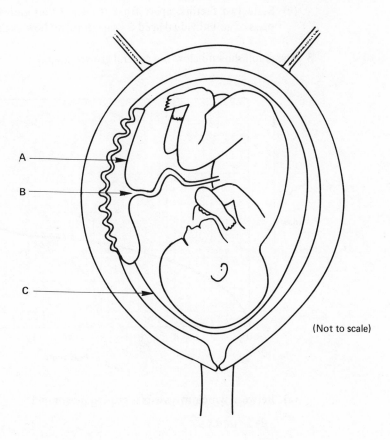

A

B

C

(Not to scale)

(i) Identify the parts labelled **A–C**. (3)
(ii) State concisely the advantages of the foetus being surrounded by a watery fluid during development. (3)
(b) (i) Name **two** nutrients which pass from the maternal blood supply to the foetus. (2)
(ii) Describe how oxygen in the maternal blood passes into the foetal blood supply. (7)
(c) Which hormone is produced by part **A** during the latter six months of pregnancy? What is its role? (3)
(d) Describe **two** changes which must occur to the circulatory system of the foetus soon after birth. (2)

(AEB, 1983)

2 The table shows the average masses of boys and girls in a group of children from birth to the age of 16 years.

Boys		Girls	
Age (years)	Mass (kg)	Age (years)	Mass (kg)
0 birth	3.5	0 birth	3.3
1	10.0	1	9.5
4	16.5	4	16.5
8	27.5	8	25.5
12	38.0	12	40.0
16	59.0	16	53.5

(a) Plot these data, on the graph paper provided, using a single set of axes. **(5)**

(b) *Using information shown on the graph.*
   (i) In between which years does the greatest increase in mass occur in boys? **(1)**
   (ii) After the age of 4 years, when are the average masses of boys and girls the same?
   (2)

(c) State **two** factors, apart from the sex of an individual, which could determine the mass of an individual aged 6 years. Explain how each exerts its influence. **(4)**

(AEB, 1983)

3 The graph shown below is a general growth graph.

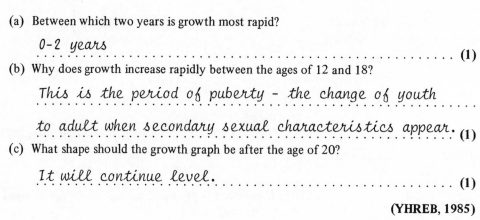

(a) Between which two years is growth most rapid?

0-2 years
.... **(1)**

(b) Why does growth increase rapidly between the ages of 12 and 18?

This is the period of puberty - the change of youth

to adult when secondary sexual characteristics appear. **(1)**

(c) What shape should the growth graph be after the age of 20?

It will continue level. .... **(1)**

(YHREB, 1985)

210

4 (a) The diagram shows a uterus during pregnancy.

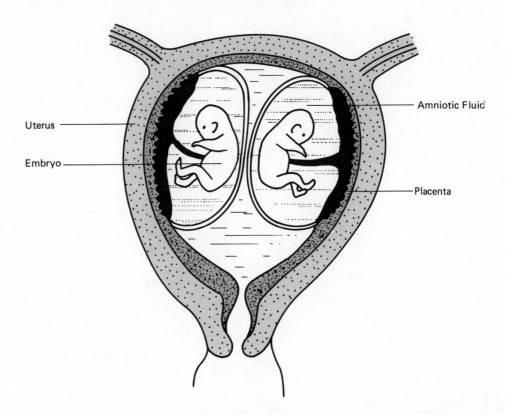

Uterus

Embryo

Amniotic Fluid

Placenta

(i) Give **two** functions of the amniotic fluid during development of the embryo.
(2)

It buoys up the foetus, allowing it to grow. It evens out the pressure changes resulting from body movement of the mother.

(ii) The placenta contains blood vessels of both the mother and the embryo. Explain the importance of this arrangement. (5)

The maternal blood vessels bring oxygen, glucose and amino acids to the placenta, and these are passed into the foetal blood vessels. The excretory products of the foetus, carbon dioxide and urea, are brought to the placenta by foetal blood vessels and exchanged to the mother's blood vessels. There is no direct connection between the two sets of vessels.

(iii) Give **two** features, shown in the diagram, which indicate that the two embryos in the uterus are non-identical (dissimilar) twins. (2)

They have separate placentas and umbilical cords.

They have separate foetal membranes.

(iv) Give **two** other features, not shown in the diagram, which may be seen in these babies following birth which would also show them to be non-identical twins.

(2)

They could be of different sex. They would be genetically different, showing different phenotypes.

(v) The diagram shows how identical (similar) twins are formed.

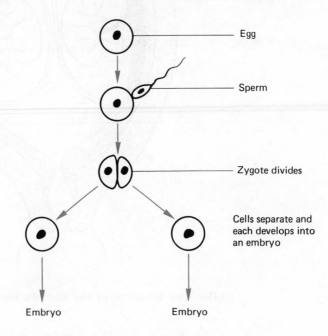

In the space below draw a similar diagram to show how non-identical twins are formed.

(3)

(SREB)

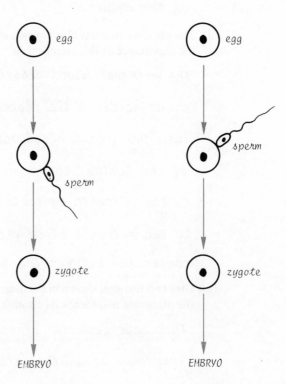

5 The figure is a diagram of part of the female body based on a vertical section.

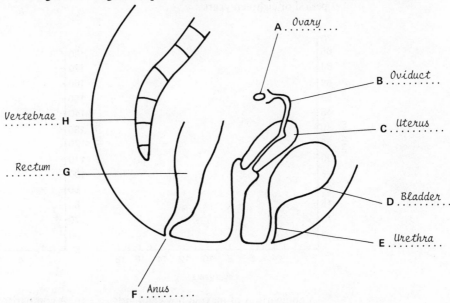

A. *Ovary*

B. *Oviduct*

Vertebrae H

C. *Uterus*

Rectum G

D. *Bladder*

E. *Urethra*

F. *Anus*

(a) Label the parts **A–H** on the diagram. **(8)**

(b) What is the name given to the lining of **C**? *Endometrium* **(1)**

(c) Which hormone from the pituitary gland affects organ **A**? *FSH* **(1)**

(OLE)

6 (a) State for each of the hormones oestrogen and progesterone its site of production and **one** effect its presence has on the uterus.

(i) Oestrogen.

Site *Ovaries*

Effect *Initiates thickening of uterus lining each month.*

(ii) Progesterone **(2)**

Site *Corpus luteum*

Effect *Causes uterus lining to thicken and vascular supplies to increase.* **(2)**

(b) On the graph draw labelled lines to represent how the amounts of each of the **two** hormones varies during the menstrual cycle of a non-pregnant woman. **(2)**

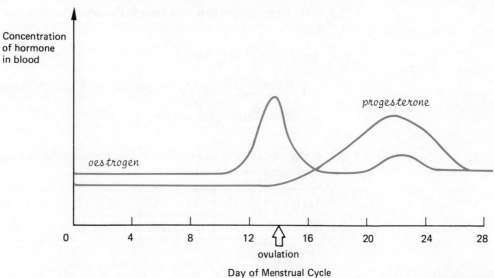

Concentration of hormone in blood

oestrogen

progesterone

0   4   8   12   16   20   24   28

ovulation

Day of Menstrual Cycle

(AEB, 1984)

213

7 The figures below represent the increases in mass and height of boys and girls over a period of eighteen years.

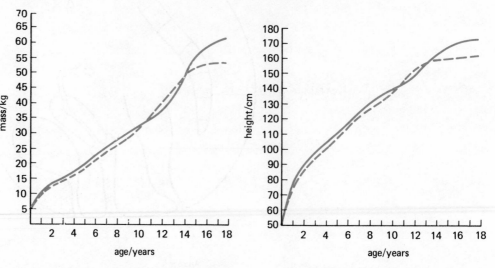

A comparison of mass in boys and girls   A comparison of height in boys and girls

(a) Which line (solid or dotted) in each figure represents girls?

> *Dotted line*

Give reasons for your answer.

> *In both curves for mass and height they overtake boys (solid line) at about the age of 10/11 years. By the age of 13/14 years, however, the boys have overtaken the girls.* **(2)**

(b) State **three** changes that take place in the male body between the ages of 13 and 18 years.

(i) *Growth of hair on pubis, under arms, on chest*

(ii) *Deepening of the voice*

(iii) *Functioning of testes* **(3)**

(c) State **three** changes that take place in the female body between the ages of 11 and 17 years.

(i) *Growth of breasts*

(ii) *Growth of hair on pubis and under arms*

(iii) *Menstruation begins* **(3)**

(UCLES)

8 (a) A boy's height was measured each year on his birthday. Using the graph paper, draw a line graph from the data.

Age (years)	10	12	13	14	15	16	18	19	20	21
Height (cm)	130	135	145	150	150	155	160	170	175	175

(10)

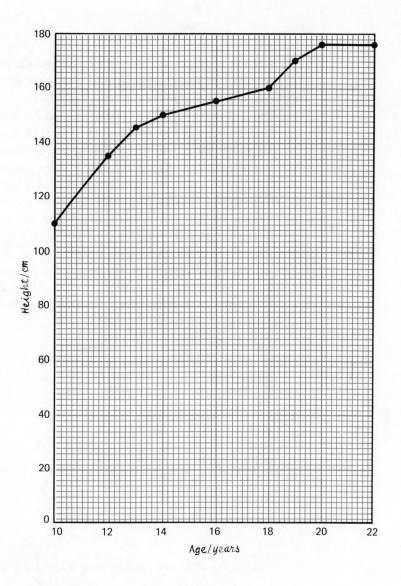

(b) Answer the following questions.
   (i) Estimate the boy's height in cm at

   A 11 years? .... *124 cm* .... **(1)**

   B 17 years? .... *158 cm* .... **(1)**

   (ii) At what age does the boy stop growing? .... *20 years* .... **(1)**
   (iii) What is the difference in the boy's height in cm between 10 years and 15 years?

   .... *20 cm* .... **(1)**

   (iv) What is the average yearly increase in cm between the ages of 12 and 15? (Show your working)

   Increase in height = 15 cm ....

   Number of years = 3 years ....

   Average increase = $\frac{15}{3} = 5$ cm year^{-1}

   .... **(4)**

   (EMREB)

215

Notes

1. In part (a) a graph should always be drawn making full use of the graph paper provided for the horizontal and vertical axes. Note that the units are written with a solidus (/), e.g.   /years.
2. Each axis should begin at 0 but in this case, with the supplied data beginning at 10 years, it would be acceptable to start at this age.
3. Actual points should be marked with a × or ○ and never by a dot only.
4. When reading from the curve, use a ruler along the correct line, e.g. in (b) (i) A place the ruler vertically along the 11 year line and read off the height on the vertical axis where the ruler cuts the curve.
5. In part (b) the correct units must always be included. Do not simply write down the answer in figures, e.g. in (b) (iv) the average yearly increase is 5 cm year^{-1} or 5 cm per year.

9 (a) (i) The chart shows how the lining of the uterus varies in thickness during the menstrual cycle. The first month is a normal month, but during the second month fertilisation occurs. Complete the chart to show what happens to the lining in the second and third months. **(3)**

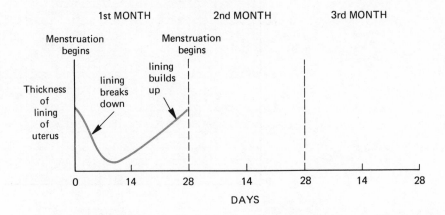

(ii) On the chart for the first month indicate with a cross when ovulation is most likely to occur. **(1)**

(iii) What is meant by fertilisation?

. . . .

. . . . **(2)**

(iv) Explain what is meant by the term menstruation.

. . . .

. . . . **(2)**

(b) (i) Label the parts **A**, **B** and **C** on the diagram of a human cheek cell. **(3)**

A . . . .

B . . . .

C . . . .

(ii) In which part of the cell would chromosomes be found?

. . . . **(1)**

(c) (i) On the diagram of the female reproductive system, mark using your own guide lines the following parts:
Using the letter W – the place where the embryo develops.
X – the place where sperms are deposited.
Y – the place where fertilisation normally occurs.
Z – the place where eggs are produced.

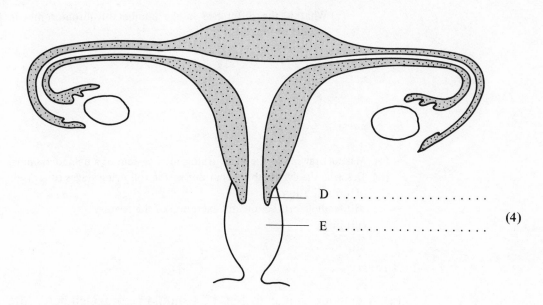

D . . . .

E . . . .   (4)

(ii) Labels parts D and E where indicated.   (2)

(d) Study the diagrams which show two sequences in the development of an egg.

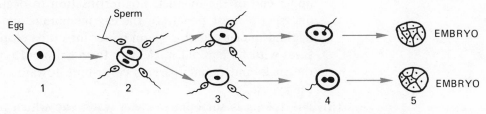

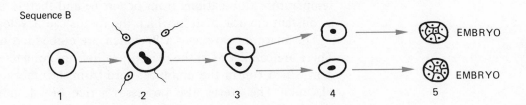

(i) In which sequence is the division of the unfertilised eggs shown?

. . . . (1)

(ii) From which sequence would non-identical twins be produced?

. . . . (1)

(iii) Write down the number of the stage which shows fertilisation in Sequence **A**.

. . . . (1)

(iv) Explain why non-identical twins may be of different sexes.

. . . .

. . . .

. . . . **(2)**

(v) What is the difference in the number of chromosomes in the ovum and the human cheek cell?

. . . . **(1)**

(WMREB)

Free-response Question

(a) Make a drawing of the male urinogenital system of a named mammal. **(7)**
(b) Describe the events that occur during the following stages of reproduction:
(i) fertilisation; **(2)**
(ii) feeding, protection and excretion of the foetus; **(7)**
(iii) birth. **(4)**

Answer

(a) A drawing similar to Fig. 12.1 should be provided here. Alternatively a lateral view as in the figure for multiple-choice questions **7–11** would be quite satisfactory.
(b) (i) Sperms are liberated into the vagina by ejaculation from the penis. By means of their tails the sperms swim up the oviducts on each side. In one oviduct the male gametes collect around the ovum, and this occurs in the upper end of the oviduct. For fertilisation to occur, the pointed end of one sperm must penetrate the egg membrane and enter the ovum. The haploid nucleus of this sperm passes into the cytoplasm of the ovum and fuses with its haploid nucleus to form a diploid zygote. In this way both the male and female parents contribute towards the genetic content of the zygote.

(ii) The foetus is surrounded by a water sac which prevents it from being damaged by the mother's movements. It also gives freedom for the embryo's own movements. The fluid in the water sac prevents sudden temperature fluctuations from occurring and therefore helps to maintain a constant environment in which the foetus can develop.

The placenta possesses villi which are embedded in the uterine wall. There are capillaries in the placenta which are connected to an artery and vein which run in the umbilical cord from the embryo's abdomen to the placenta. The uterus also possesses a rich blood supply. However, the blood vessels of the mother and placenta, although they lie close together, do not fuse. There is no mixing of the blood of mother and foetus; thus, the maternal blood pressure does not affect the foetal circulation.

The placenta prevents the entry of bacteria, toxins and proteins into the foetus from the mother. The membranes separating the mother's and placental blood vessels are very thin and, hence, dissolved oxygen, glucose, amino acids, salts and antibodies from the mother's blood pass into the embryonic capillaries, while carbon dioxide and nitrogenous waste from the embryo diffuse across into the maternal circulation in the opposite direction. In this way the placenta enables feeding and excretion to

218

occur between mother and foetus. From the foodstuffs the foetus receives, it is able to synthesise its own protoplasm and therefore grow.

(iii) At the onset of birth the uterus begins to contract rhythmically. The contractions become stronger and more frequent as birth approaches. The pubic ligaments stretch and the vaginal passage dilates sufficiently to allow the child's head to pass through. At about this time abdominal contractions reinforce those of the uterus. The water sac breaks and fluid is released through the vagina. Finally, the muscular contractions force the baby out of the womb altogether. The umbilical cord which still connects the child to the placenta is severed. Later the placenta and the remainder of the cord comes away from the uterus and is expelled as 'after-birth'.

Notes

1. The urinogenital system is drawn in ventral view. The question does not specify a particular view, so that the more simple lateral view could be drawn. It would still include all of the essential features (see Fig. 12.1).
2. The essential element of fertilisation is the fusion of the haploid nuclei. The answer for part (b) (i) must include some reference to the sex act and the deposition of sperms in the genital system of the female.
3. Part (b) (ii) commands more marks than parts (i) and (iii) and therefore requires more time and expansion of the content.
4. Part (b) (iii). The best candidates should mention the role of the hormone oxytocin in stimulating the onset of labour.

(c) Answers to Objective and Structured Questions

(i) *Multiple-choice Questions*

1. B   2. E   3. D   4. A   5. D   6. C   7. A   8. C   9. E   10. A   11. B   12. A
13. A   14. D   15. E   16. D

(ii) *Structured Questions*

1 (a) (i) A, placenta; B, umbilical cord; C, foetal membranes   (ii) The foetus is buoyed up by the amniotic fluid, enabling it to develop without any constriction. The fluid cushions it against shock and the movements of the mother's body. It permits movements of the limbs, body and head during the later stages of growth. The foetus has been observed to make breathing movements when the fluid moves in and out of the buccal cavity.

(b) (i) Glucose and amino acids

(ii) Oxygen is carried as oxyhaemoglobin in the red blood cells of maternal blood to the blood sinuses in the placenta. The oxygen is released and diffuses across the walls of the sinuses and capillaries. The oxygen combines with the haemoglobin of the red blood cells of the foetal circulation. The blood is then carried back to the foetus through the umbilical vein.

(c) Progesterone, which maintains the endometrium and inhibits ovulation.

(d) 1. The foramen ovale, an aperture connecting the right and left atria, closes up so that blood is no longer short-circuited between right and left atria.
   2. The ductus arteriosus connecting the pulmonary artery and the aorta closes so that the lungs are no longer short-circuited and breathing can commence.

2 (a) See the graph. (For notes on drawing graphs, see question **8**.)

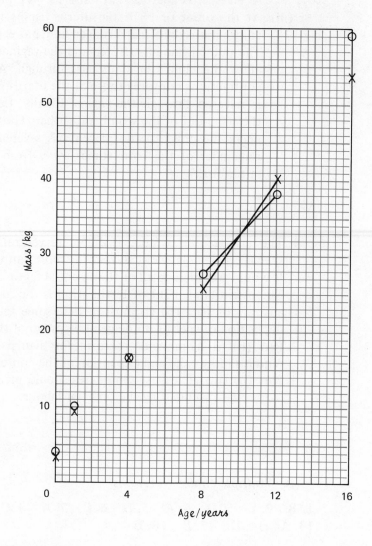

(b) (i) 12 to 16 years
   (ii) Between 10 and 11 years
(c) 1. Insufficient food, i.e. lack of protein for growth or lack of vitamin D for bone growth.
   2. Incorrect hormonal balance and lack of growth hormones, e.g. thyroxine.

Questions **3–8** have the answers supplied with the questions. Question **9** has no answers supplied. Try completing the question yourself.

13 Inheritance and Family Planning

13.1 Cell division

(a) Mitosis

See Fig. 13.1.

Mitotic division occurs in all body cells except gamete mother cells. It results in the production of new protoplasm, i.e. growth. The early divisions of the zygote and the embryo are mitotic, but as cells become specialised (i.e. muscle cells, nerve cells and blood cells), they lose the power of division. A few tissues continue to divide mitotically, e.g. bone marrow cells and the germinative layer of the skin. Each body cell has 46 chromosomes in its nucleus (**diploid number**, i.e. $2n = 46$).

(b) Meiosis

See Fig. 13.2.

Meiotic division occurs when sperm mother cells or egg mother cells divide in the testis or ovary. It results in the production of **gametes** (sperms or ova) which have half of the chomrosome number (**haploid number**, $n = 23$) found in the body cells (somatic cells). In addition, each gamete has a unique combination of chromosomes due to (a) the **random separation** of homologous pairs (bivalents) and (b) the fact that during the second meiotic division there is a **crossing-over (chiasmata)** of portions of the chromatids, so that each new chromosome may contain sections derived from both parent chromosomes.

This reduction division in meiosis is necessary to prevent duplication at every fertilisation. The fertilised ovum (**zygote**) therefore always has the diploid number (46) restored by the joining together of the nuclei of sperm and ovum. Each body nucleus has the following numbers of chromosomes:

22 pairs of **autosomes** + 1 pair of **sex chromosomes** = 23 pairs
in the male ♂: 22 pairs of autosomes + 1 pair of sex chromosomes XY
in the female ♀: 22 pairs of autosomes + 1 pair of sex chromosomes XX

13.2 Inheritance: definitions of terms used

Allele (allelomorph): one of a pair (or more) of alternative forms of a gene.
Bivalent: a pair of homologous chromosomes, situated together during meiosis.
Chromatid: one half of the chromosome, produced when the chromosome splits longitudinally during cell division.

1

EARLY PROPHASE
During the interphase the cell carries out everyday activities. In prophase, the chromosomes begin to shorten and the nucleolus begins to shrink.

2

LATE PROPHASE
Chromosomes shorten further and duplicate themselves; each one consists of a pair of chromatids joined at the centromere. Centrioles move apart, nucleolus disappears.

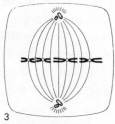

3

METAPHASE
Nuclear membrane breaks down, spindle forms, chromatid pairs line up on equator of cell. (Single chromosome per spindle.)

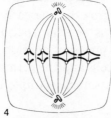

4

ANAPHASE
Chromatid pairs separate and move to opposite ends of cell.

5

EARLY TELOPHASE
Chromatids reach destination and reform into chromosomes. Cell begins to constrict at equator.

6

LATE TELOPHASE
Constriction completed and nuclear membrane and nucleolus reform in each cell. The chromosomes become invisible threads and each cell enters interphase.

Fig. 13.1 The main stages of mitosis in a generalised animal cell

1

EARLY PROPHASE   1
Homologous chromosomes appear in nucleus.

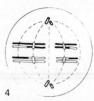

2

MIDDLE PROPHASE   1
Homologous chromosomes pair up, then split into chromatids.

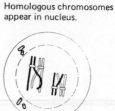

3

LATE PROPHASE   1
Chromatids of homologous chromosomes cross over each other.

4

METAPHASE   1
Homologous chromosomes arrange themselves on equator of spindle. Segments of crossed chromatids have exchanged by breakage and subsequent rejoining.

5

ANAPHASE   1
Homologous chromosomes separate.

6

TELOPHASE   1
Two new cells form, each with half the number of chromosomes of the original cell. Nuclear membrane may not reform, as metaphase 2 may follow immediately.

7

METAPHASE 2
A new spindle is formed in each new cell and two chromosomes line up on each equator.

8

ANAPHASE   2
Chromatids of the two chromosomes in each new cell separate (as in mitosis).

9

TELOPHASE   2
Four new cells form, each with half the number of chromosomes compared with the original cell. The composition of the chromosomes is altered.

Fig. 13.2 The main stages of meiosis, involving two pairs of homologous chromosomes

Chromosome: a thread-like structure, bearing genes and located in the nucleus.

Diploid: describes a cell that has two sets of homologous chromosomes.

Dominant: describes an allele whose effect is seen in the phenotype of the heterozygote, in spite of the presence of an alternative allele.

Gene: a unit of hereditary material, located on the chromosome.

Genotype: a description of an organism in terms of its genes.

Haploid: describes a cell with a single set of chromosomes.

Heterozygote: an organism which has two **different** alleles of the same gene.

Homozygote: an organism which has two **identical** alleles of the same gene.

Phenotype: a description of an organism in terms of what can be seen or measured.

Recessive: describes an allele whose effect is not seen in the heterozygote, because of the presence of a dominant allele of the same gene.

13.3 Monohybrid inheritance (single-factor inheritance)

The following illustrates the Mendelian principles of inheritance (name derived from Gregor **Mendel**, an Austrian monk) in Man.

Example 1

Let **T** represent the allele for rolling the tongue. Let t represent the allele for non-ability to roll the tongue.

Parents (P) phenotype tongue rolling × non-tongue rolling
   genotype   **TT** ×   tt
   gametes **T** **T** t   t

First filial generation (F1) genotype (all) **Tt**
          phenotype (all) tongue rolling

This illustrates **Mendel's first law** (principle of segregation): 'The characteristics of an organism are determined by internal factors which operate in pairs. Only one of a pair of such factors can be represented in a single gamete'.

If the F1 generation of the cross in Example 1 is mated with another heterozygote, the results are as follows:

Example 2

P phenotype   tongue rolling × tongue rolling
 genotype   **Tt** × **Tt**
 gametes   **T** t **T**   t

Second filial generation (F2) TT Tt tT tt genotypes

   phenotypes 3 tongue rolling 1 non-tongue rolling

Note:

1. Tongue rolling is dominant to non-tongue rolling, which is recessive.
2. In the second cross Mendel's first law still applies, except that there are two types of gamete for each parent.
3. Crossing two heterozygotes always gives a 3 : 1 ratio in the phenotypes and a 1 : 2 : 1 ratio in the genotypes.

Example 3

Inheritance of the rhesus factor Rh+ and Rh−. Rh+ is dominant to Rh−: 85% of the population are Rh+.

Let **R** represent the allele for Rh+. Let **r** represent the allele for Rh−.

P	phenotype	Rh+	×	Rh−
	genotype	RR	×	rr
	gametes	R	×	r
F1	genotype		Rr	
	phenotype		(all) Rh+	

All of the offspring are heterozygous and Rh+. In a marriage of two Rh+ parents who are heterozygous the resulting children of the couple will be Rh+ and Rh− in the ratio 3:1.

Note that all of these ratios apply to **theoretical results** when **large numbers of families** are considered. In Man with a limited number of children to each marriage these ratios will not necessarily apply to a single marriage. For example, in the second mating above (rhesus factor) it is possible to have four children all of whom are Rh+, but it is also possible to have only one child who is Rh−.

(a) Incomplete Dominance (Co-dominance)

Some alleles do not show complete dominance. Thus, in some plants there would be an intermediate flower colour for the heterozygote, i.e.

$$\text{red} \quad \times \quad \text{white}$$
$$|$$
$$\text{pink}$$

This phenomenon occurs in the inheritance of human blood groups (see Chapter 5). This is governed by three alleles: G^A, G^B and G (or G^O). G^A and G^B are both dominant to G, but G^A **and G^B are not dominant** to each other, i.e. the **two genes are co-dominant**.

Thus,
blood group A has genotypes $G^A G^A$ or $G^A G$;
blood group B has genotypes $G^B G^B$ or $G^B G$;
blood group AB has genotype $G^A G^B$ (co-dominance);
blood group O has genotype GG.

See structured question 7.

13.4   Sex linkage

When the genes controlling a certain character are located on the sex chromosomes, the character is inherited along with the sex of the individual. Such characters are described as sex-linked. The Y chromosome in Man is thought to contain very few genes, if any at all. Thus, a gene carried on the X chromosome in the male need not have a corresponding gene on the Y chromosome.

(a) Haemophilia

The inheritance of haemophilia is one of the best-known examples of sex-linked inheritance. Female haemophiliacs are rare, since the disease is carried on the X

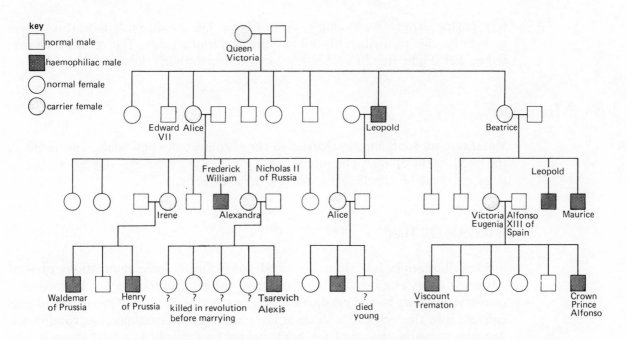

Fig. 13.3 Queen Victoria's family tree, showing inheritance of the recessive sex-linked gene for haemophilia

chromosome. However, a woman with the recessive gene for haemophilia on one of her X chromosomes will act as a 'carrier' for the disease. Queen Victoria was such a carrier (see Fig. 13.3).

Example 4

Let the gene for haemophilia be **r**. Let the gene for normal blood clotting be **R**. Let X represent the X chromosome. Let Y represent the Y chromosome.

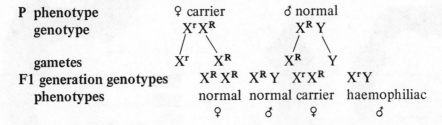

Note:
    1. The four genotypes of this cross represent the two possible phenotypes for each sex: i.e. half of the sons are haemophiliacs and half are normal; half of the daughters are 'carriers' and half are normal.
    2. Consider the results of the following cross:

$$X^r X^R \times X^r Y$$

The F1 generation includes $X^r X^r$, a female homozygous for haemophilia. The daughter does not grow into a haemophiliac, since the foetus aborts.

(b) Red–Green Colour-blindness

About 8% of the male population suffers from red–green colour-blindness. It is particularly important that men recruited into the Navy or to be trained as railway engine drivers, for example, do not have this condition. It is vital for these men to be able to distinguish red and green warning lights. The condition is sex-linked and is inherited in a similar fashion to haemophilia, through 'carriers'.

13.5 Mutations

Mutations are spontaneous changes in the chromosomes and genes, and result in the production of new phenotypes. They are often recessive and lethal, but in some cases can be beneficial (selectively advantageous).

(a) Sickle-cell Trait

Sickle-cell anaemia is a common blood condition in some populations in West Africa. It arises as a result of a mutation of the gene responsible for haemoglobin formation leading to the production of a new haemoglobin S. The red blood cells collapse into the shape of a sickle at low oxygen concentrations. The condition is due to a recessive gene, and the homozygous form results in a fatal anaemia. The heterozygous form, however, has some normal haemoglobin and some haemoglobin S, and the individual suffers from a milder form of anaemia. This incomplete dominance carries an advantage, in that anyone who is a heterozygote has a high resistance to malaria, which can be a fatal disease in West Africa.

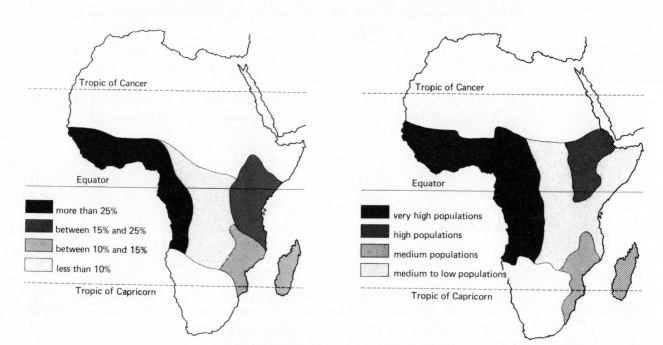

Fig. 13.4 Map showing the distribution frequency of people carrying the gene for sickle cell haemoglobin

Fig. 13.5 Map showing the distribution frequency of *Anopheles* mosquitoes, which act as vectors for the organisms carrying malaria

(b) Down's Syndrome (Mongolism)

Down's syndrome is a mutation occurring about 16 times in every 10 000 births. It is caused by an extra chromosome of the 21st pair, giving a total of 47 instead of 46.

13.6 Family limitation

Richer, better-fed people (in developed countries) tend to have smaller families — birth rate about 25 per 1000 of the population. Poor, underfed people (in underdeveloped countries) have higher birth rates of about 55 per 1000 of the population. Better feeding, however, tends to increase fertility, so that the explanation of the above figures is probably social rather than biological. There are a number of social factors operating in wealthy societies:

1. Growth of cities is associated with lower fertility.
2. Higher standards of living correlate with lower fertility.
3. Employment of women means a drop in the birth rate.
4. Economic depressions cut birth rates.
5. Availability of contraceptive practices to use and limit family size.

13.7 Methods of contraception

1. The **natural or rhythm method** means restricting intercourse to the non-fertile period of the woman, lasting from a few days after ovulation until a few days before the next ovulation. The method depends upon a regular menstrual cycle; it can involve taking daily body temperatures and thus determining the exact date of ovulation (see Fig. 13.6). It is the least reliable method.
2. **Mechanical methods**
   (a) The placing of a barrier between egg and sperm by means of rubber. For male use, the condom or 'sheath'. For female use, the diaphragm or 'Dutch cap', a rubber dome designed to fit into the end of the vagina over the cervix (see Fig. 13.7).

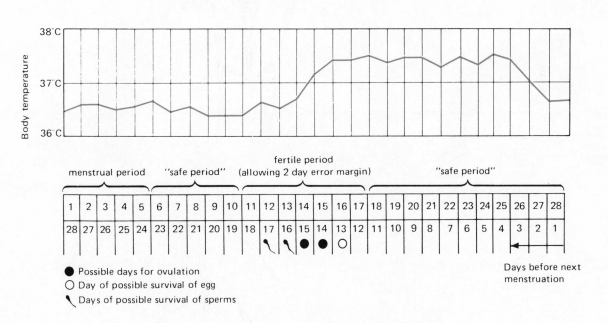

Fig. 13.6   The rhythm method of family planning

   (b) The intra-uterine device (IUD) is a coil of plastic placed inside the uterus to prevent implantation (see Fig. 13.8).
3. **Chemical methods**
   (a) Spermicidal jelly placed in the vagina to kill sperms.

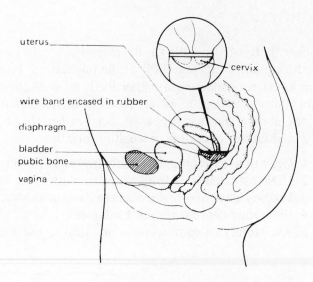

Fig. 13.7 Contraceptive diaphragm in position

(b) 'Birth control' or hormonal pills taken orally (one per day) from the 5th to the 25th day of the menstrual cycle. The pills contain a mixture of oestrogen and progesterone-like hormones which act on the feedback mechanisms of the endocrine system (see Chapter 8).

4. **Surgical methods** involve vasectomy for men and tubal ligation for women (see Figs. 13.9 and 13.10).

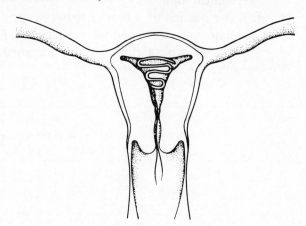

Fig. 13.8   IUD in position

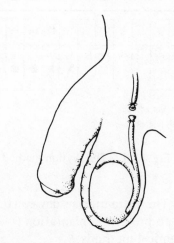

Fig. 13.9   Vasectomy

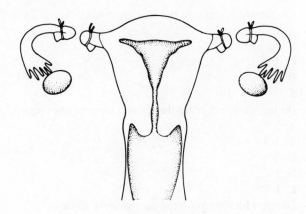

Fig. 13.10 Tubal ligation

See Table 13.1 for a comparison of the effectiveness of different methods, classified in terms of the number of pregnancies per 100 women in one year.

Table 13.1 Comparative effectiveness of birth control

Method	Pregnancies per 100 woman-years
No contraception practised	45
Rhythm (safe period)	25
Spermicidal cream only	20
Withdrawal of penis before ejaculation	18
Condom	14
Diaphragm with spermicide	12
IUD	2
Hormonal pill	0.5
Sterilisation	0

13.8 Questions and answers

(a) Multiple-choice Questions

Questions **1–4** refer to the stages of mitosis listed below.
   1 metaphase
   2 prophase
   3 interphase
   4 telophase
   5 anaphase

   1 During which stage does all cell growth take place?
   A 1
   B 2
   C 3
   D 4
   E 5

2 During which stage do the chromosomes come to lie on the equatorial plate?

  A 1
  B 2
  C 4
  D 5
  E 3

3 During which stage are the nuclear membranes formed?

  A 2
  B 3
  C 4
  D 5
  E 1

4 During which stage do the centromeres divide?

  A 2
  B 3
  C 4
  D 5
  E 1

5 In meiosis, how many chromatids are found in each bivalent?

  A 1
  B 2
  C 3
  D 4
  E 5

6 On full completion of meiosis in the male, how many spermatozoa are produced from a single spermatogonium?

  A 1
  B 2
  C 3
  D 4
  E 5

7 On full completion of meiosis in the female, how many egg cells are produced from a single oogonium?

  A 1
  B 2
  C 3
  D 4
  E 5

8 In humans, the gene for the ability to roll the tongue (**R**) is dominant to the gene for the inability to roll the tongue (**r**). Two brothers, whose parents were both tongue-rollers, also have the ability to roll their tongues. One of the grandfathers could not roll his tongue. What are the genotypes of the two brothers?

  A **RR** and **RR**
  B **Rr** and **Rr**
  C **RR** and **Rr**
  D **RR** and **rr**
  E impossible to tell from information given

9 What would be the genotype of the two brothers in question 8 above if neither grandfather could roll his tongue?

  A **RR** and **RR**
  B **Rr** and **Rr**
  C **RR** and **Rr**
  D **Rr** and **rr**
  E impossible to tell from information given

10 If a man's mother is homozygous for blood group **A** and his father is homozygous for blood group **B**, what are the chances of his own blood group being **A**?

  A 0%
  B $33\frac{1}{3}$%

C 50%

D $66\frac{2}{3}$%

E 100%

11 A man, homozygous for blood group A, marries a woman whose genotype is G^BG. What is the probability of their first child having blood of group A?

    A 0%

    B $33\frac{1}{3}$%

    C 50%

    D $66\frac{2}{3}$%

    E 100%

12 A man with blood group AB marries a woman with blood group O. What is the probability of their first child having blood group AB?

    A 0%

    B $33\frac{1}{3}$%

    C 50%

    D $66\frac{2}{3}$%

    E 100%

13 What is the sex chromosome content of the human egg cell?

    A XXY

    B XY

    C XX

    D Y

    E X

14 Haemophilia is caused by

    A a dominant gene carried on the X chromosome.

    B a dominant gene carried on the Y chromosome.

    C a recessive gene carried on an autosome.

    D a recessive gene carried on the X chromosome.

    E a recessive gene carried on the Y chromosome.

Questions **15–18** refer to the list below.

1 allele
2 bivalent
3 chromatid
4 diploid
5 genotype
6 haploid
7 heterozygote
8 homozygote
9 phenotype

15 An organism which has two different alleles of the same gene.

    A 2

    B 4

    C 7

    D 8

    E 9

16 A description of an organism in terms of what can be seen or measured.

    A 2

    B 5

    C 7

    D 8

    E 9

17 Describes a cell that has two sets of homologous chromosomes.

    A 2

    B 4

    C 5

    D 7

    E 9

18 A description of an organism in terms of certain of its genes.

   A 1
   B 2
   C 5
   D 8
   E 9

19 Which one of the following is the best definition of a gene?

   A a group of chromosomes in the nucleus
   B a portion of a chromosome responsible for several characteristics
   C a part of a single chromosome in the nucleus
   D a part of a chromosome responsible for producing one characteristic
   E a factor responsible for producing a characteristic

20 Which one of the following is caused by the presence of an extra chromosome in the body nucleus?

   A brown eyes
   B sickle-cell anaemia
   C haemophilia
   D tongue rolling
   E Down syndrome (mongolism)

Questions **21–23** refer to the figure below.

T = tongue rolling   **t** = inability to tongue roll

Parental generation   **TT**   ×   **tt**

Gametes   **T**   **T**   **t**   **t**
F1 generation   (all) **Tt**

Second marriage   **Tt**   ×   **Tt**

		sperms	
		T	**t**
eggs	**T**	**TT** ①	**Tt** ②
	t	**tT** ③	**tt** ④

F2 generation

21 Which of the second filial generation is homozygous, unable to roll the tongue genotype?

   A 1
   B 2
   C 2 and 3
   D 3
   E 4

22 Which of the second filial generation is a tongue-rolling phenotype?

   A 1
   B 4
   C 1, 2 and 3
   D 1 and 2
   E 2 and 3

23 Which of the second filial generation is heterozygous genotype?

   A 1
   B 1, 2 and 3
   C 2 and 3
   D 1 and 3
   E 4

24 Colour-blindness is a sex-linked recessive in Man. A woman with normal sight marries a colour-blind man. Assuming that they have several children, which of the following chances will be most correct?

   A None of the daughters will be colour-blind but the sons will.
   B None of the sons will be colour-blind but the daughters will.

C None of the sons or daughters will be colour-blind.

D All of the daughters will be colour-blind.

E All of the sons will be colour-blind.

(b) **Structured Questions**

1 Immediately after birth a set of triplets, two of whom were known to be identical twins, were separated and reared under very different conditions. The table below shows data for the triplets at the age of 20.

	Mary	*June*	*Elizabeth*
Height in m	1.78	1.78	1.74
Weight in kg	78	80	86
Blood group	O	AB	O
I.Q.	135	140	125

(a) Which two girls were identical twins? (one line) **(1)**

(b) Which characteristic shown in the table enabled you to answer (a)? (one line) **(1)**

(c) The three girls had got together to try to find their original parents, and had narrowed the search to four possible couples:

> Couple 1 – one had type A blood, the other type B;
>
> Couple 2 – both had type AB blood;
>
> Couple 3 – one had type A blood, the other type AB;
>
> Couple 4 – one had type O blood, the other type AB.

(i) Which couple do you think were the parents? (one line) **(1)**

(ii) Give reasons for your answer to (c) (i). (two lines) **(2)**

(d) Both parents were known to have brown eyes; Mary has blue eyes. Assuming that the brown allele (**B**) is dominant to the blue allele (**b**), what were the genotypes of (i) the mother and (ii) the father?

(i) mother (one line) (ii) father (one line) **(2)**

(e) What is the probability that June had blue eyes? (one line) **(1)**

(L)

2 (b) Some persons can roll the tongue into a tube-like structure. Others cannot do this action. The ability to roll the tongue is dominant and is controlled by the allele **T**. The inability to roll the tongue is recessive and is controlled by the allele **t**. The diagram shows a family pedigree for this characteristic. The genotype of each person is also shown.

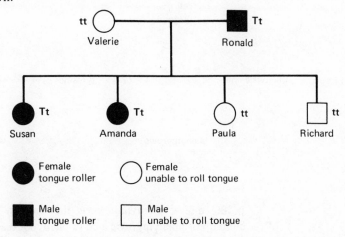

(i) Susan is shown as having the genotype **Tt**. Explain how it is possible to know this, using information from the diagram. (one line) **(2)**

(ii) Richard is homozygous for the characteristic of tongue rolling. Explain the meaning of the term *homozygous*. (one line) **(1)**

[Part question] **(SREB)**

233

3 (a) The diagram represents a family pedigree showing the inheritance of *phenylketonuria*, which is controlled by a single pair of alleles, **A** and **a**.

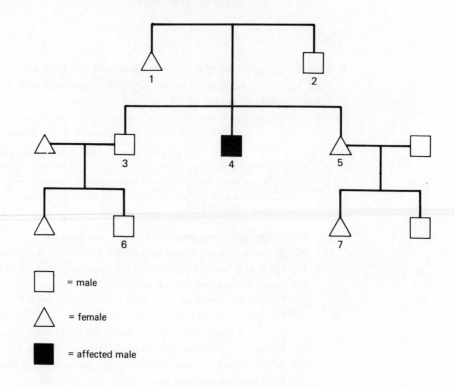

□ = male

△ = female

■ = affected male

(i) Is the allele for phenylketonuria dominant or recessive? **(1)**

(ii) State with reasons the genotype of each of individuals **1, 2** and **4**. **(4)**

(iii) What are the possible genotypes of each of individuals **3** and **5**? **(2)**

(iv) Why would it be genetically inadvisable for cousins **6** and **7** to marry and have children? **(3)**

(b) The diagram shows a simplified sequence of chemical reactions which occur in the body.

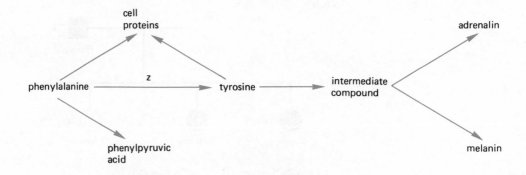

In the disease *phenylketonuria* the reaction labelled z does not take place and a build-up of phenylpyruvic acid (a phenylketo-acid) occurs.

(i) What is a gene mutation? **(2)**

(ii) A gene mutation has given rise to the inability to carry out reaction z. Explain this statement. **(5)**

(iii) Infants suffering from *phenylketonuria* often have fair hair and fair complexion. Explain briefly, using information from the diagram, why this is so. **(3)**

(AEB, 1984)

4 A blood disorder leading to a form of anaemia has been found to be inherited. Investigation of the medical records of a particular family yielded the following information, though in some cases the records gave no information regarding the presence or absence of symptoms.

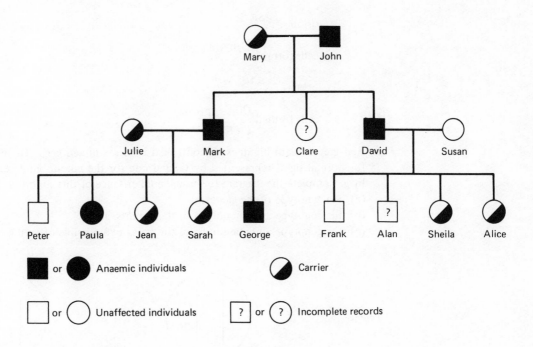

(i) What is meant by the term 'carrier'?
(ii) In your answers to *each* of the following questions explain your reasoning clearly.
   (a) What is the likely genotype of Alan?
   (b) What is the probability that Clare is a carrier?
   (c) If Frank was to marry a woman with the same genotype as Sarah what proportion of any sons might be expected to be anaemic?   **(7)**
                                                                                        (NISEC)

5 (a) Briefly describe the results which occur when a cell nucleus divides by mitosis.

Two new nuclei are formed. Each nucleus has the same number of chromosomes as the original nucleus. **(2)**

(b) Why is it particularly important that mitosis should take place frequently in the following structures?

   (i) Skin epidermis *The epidermal cells are continually dividing, producing new cells which are gradually pushed outwards. The outermost cells die.*

   (ii) Bone marrow *The bone marrow continually produces new red blood cells. They have a very limited life and their numbers must be replaced.* **(4)**

(c) Under ideal conditions a cell can grow and divide in 30 minutes. Starting with one cell, how many cells would be formed in such ideal conditions in a period of 3 hours?

   32 **(1)**

(d) (i) Apart from mitosis, what other kind of nuclear division takes place?

Meiosis .... **(1)**

(ii) State *two* organs in which this nuclear division takes place. In each case, name the cells which are formed.

organ *Testis* ....

cells formed *Sperms* ....

organ *Ovary* ....

cells formed *Ova (eggs)* .... **(4)**

(L)

6 Red–green colour-blindness is controlled by a sex-linked gene. There are two alleles, one for normal sight, represented by **G**, and one for the colour-blind condition, represented by **g**. Complete the diagram to show the inheritance of this condition by inserting
(a) the genotype of the gametes,
(b) the genotype and phenotype of the parents, and
(c) the genotype and phenotype of the other two possible offspring.

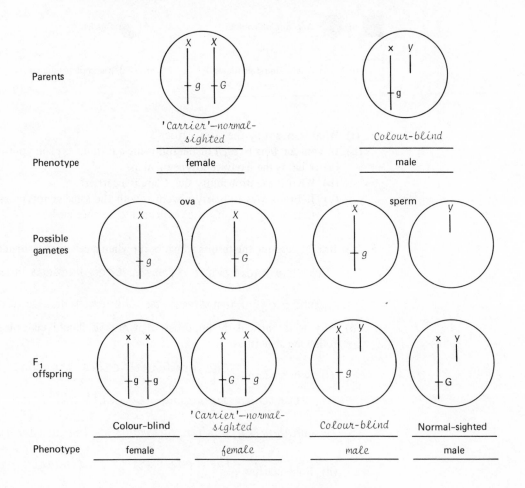

(4)

(AEB, 1983)

Notes

The unusual mating of a colour-blind man with a 'carrier' woman can be deduced from the stated first offspring in the F1. Note that it is a double recessive for the

236

colour-blind condition, with two **gg**. This can only be achieved if the mother is heterozygous and the father colour-blind.

7 The figure below shows a diagram showing three generations of a family tree, together with the blood groups, on the **ABO** system, of the family members.

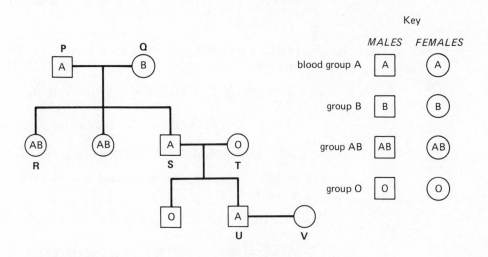

Blood groups are inherited from three genes:
Group A from gene G^A; group B from gene G^B; group AB from genes G^AG^B; group O from gene G. Genes G^A and G^B are co-dominant (show incomplete dominance), while gene **G** is recessive. Study the diagram, where necessary, and answer the following questions.

(a) What is the genotype (genes present) of person **R**?

$G^A \, G^B$

(b) (i) What is the genotype of person **S**?

$G^A \, G$

(ii) Give reasons for your answer to (i) above.

Genotype of P = G^AG^A OR G^AG; genotype of Q = G^BG^B

OR G^BG; S must be G^AG, since G^AG^A not possible in the

cross P × Q.

(iii) What is the genotype of person **Q**?

G^BG

(c) (i) What is the genotype of person **T**?

GG

(ii) Explain how it is possible for parents **S** and **T** to produce a child with blood group O.

S = G^AG × T = GG

half children are G^AG and half are GG, i.e. half are

Group A and half are Group O

(d) Parents **U** and **V** produce four children, each with a different blood group. Explain how this is possible.

V must be group B genotype $G^B G$.

Then parents $G^A G \times G^B G$

gametes G^A G $\times$ G^B G

Offspring $G^A G^B$ (group AB); $G^A G$ (group A); $G^B G$ (group B); GG (group O)

(e) (i) Explain what will happen when blood from person **T** is mixed with serum from person **S**.

T = group O (GG)   S = group A ($G^A G$)

Group O red corpuscles in serum with antibody b will not clump together (agglutinate).

(ii) Explain what will happen when blood from person **S** is mixed with serum from person **T**.

Group A red corpuscles in serum with antibodies at will clump together (agglutinate).

(f) From which blood groups can person **Q** safely receive a transfusion?

Group B and group O .... **(12)**

(UCLES)

Free-response Question

(a) Name the organs in which gamete production occurs in mammals. **(2)**
(b) Approximately equal numbers of males and females are born. Suggust reasons to explain why this is so. **(6)**
(c) Why is it that blood groups A and B in Man can exist in both the heterozygous and homozygous conditions, yet group O can only exist in the homozygous state? **(4)**
(d) What are the possible blood groups likely to be inherited by children born to a group A father and a group B mother? Fully explain your reasoning. **(8)**

Answer

(a) Gametes are produced in the ovaries of the female mammal and the testes of the male.
(b) Men and women possess twenty-three pairs of chromosomes. Of these, one pair, the sex chromosomes, determine the sex of the individual. In males the sex chromosomes are dissimilar from each other and are called the X and Y chromosomes, Y being very short and genetically empty. In females, the sex chromosomes are identical with each other and are called X and X.

Sperms produced by a male will contain either the X or the Y chromosome in their chromosome content, but all ova produced by females will contain the X sex chromosome only.

As fusion of the gametes is a random process, it follows that the female X chromosome of an ovum could join with either a sperm containing a Y chromosome or one containing an X chromosome, as shown in the diagram.

$$\begin{array}{c} \text{sperm} \\ \begin{array}{c|c|c} & X & Y \\ \hline X & XX & XY \\ \hline X & XX & XY \end{array} \end{array}$$

ova

Therefore, it can be seen that males and females are likely to be formed in equal numbers.

(c) The blood group of an individual is determined by genes at a single spot (locus) on the chromosome. There are three alternative genes (alleles), **A**, **B** and **O**. Any person can only possess two of these genes, one per homologous chromosome. **A** and **B** are equally dominant, while **O** is recessive to both of them. Therefore, it is possible to have blood group A as **AA** (homozygous) or **AO** (heterozygous), and group B as **BB** or **BO**. However, group **O** can only exist phenotypically as **OO**, the homozygous recessive state.

(d) The group A father could possess the genotype **AA** or **AO**. Likewise, the group B mother may be **BB** or **BO**. Therefore, there are four possible crosses between mother and father, as follows:

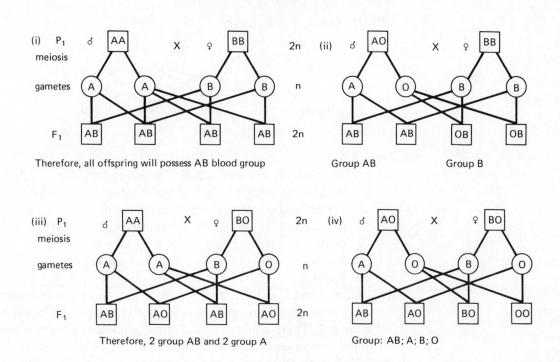

Notes

1. Explanation (b) requires statements regarding the number of chromosomes and the way they behave during gamete formation. Furthermore, this question can only be fully answered in terms of X and Y chromosomes and their segregation in male and female gametes.

2. Section (c) requires knowledge of the three genes determining blood groups and the important fact that genes **A** and **B** are equally dominant.

3. Section (d) has been answered in the form of a genetic diagram. In this case there is no need to define genes, because they refer to blood groups bearing the similar letters. Normally, in a genetic question the key for genes must be given first, showing to which phenotype they belong, e.g. let **Y** indicate yellow seeds and **y** indicate green seeds.

239

Notice there are four possible crosses. For the single cross mentioned always ensure that your answer shows every possible cross.

(c) Answers to Objective and Structured Questions

(i) *Multiple-choice Questions*

1. C 2. A 3. C 4. A 5. D 6. D 7. A 8. E 9. E 10. A 11. C 12. A 13. E
14. D   15. C   16. E   17. B   18. C   19. D 20. E 21. E 22. C 23. C
24. C

(ii) *Structured Questions*

1 (a) Mary and Elizabeth   (b) Each blood group O   (c) (i) Couple 1   (ii) Only couple with genotypes $G^A G^O$ and $G^B G^O$ and this mating could give off-spring $G^A G^B$, $G^A G^O$, $G^B G^O$, $G^O G^O$

(d) (i) Mother genotype **Bb**

(ii) Father genotype **Bb**

(e) 1 chance in 4 or 25% probability

2 (b) (i)

| Parents | Valerie | X | Ronald |

or Susan shown as a female tongue roller (black circle). She could only be heterozygous because of the mother who was homozygous recessive.

(Since the question only supplies one line for the answer, it must be the second of the above alternatives. If further space were allowed, it could be in terms of a genetic diagram.)

(ii) Richard receives like genes from each parent.

3 (a) (i) Recessive   (ii) 1, **Aa**; 2, **Aa**; 4, **aa**;   (iii) 3 and 5 could be either **AA** or **Aa**.   (iv) If 3 is **Aa** and 5 is **Aa**, then there is a recessive gene in one of the parents for each of 6 and 7. That recessive gene could be handed on to 6 and 7. Their children could show a homozygous recessive phenotype if these two genes came together, i.e. the disease phenylketonuria.

(b) (i) A gene mutation is the sudden appearance, generally in a known proportion in the population, of a new characteristic. It is caused by a change in the structure of DNA at a single locus on a chromosome.

(ii) The gene–enzyme hypothesis indicates that a gene may show itself as an enzyme. In this case the mutation has changed the enzyme so that it no longer catalyses the change from phenylalanine to tyrosine.

(iii) The correct sequence of amino acids and intermediate compounds would result in the production of melanin, a dark pigment in the skin. The mutation has stopped this sequence, so that the infants will not have melanin in their skin, i.e. they will have fair hair and a fair complexion.

4 (i) In this case the female is heterozygous and carries the recessive gene, which is masked by the dominant gene in the phenotype.

(ii) (a) Let the genes be carried on the X chromosome, the gene for normal blood condition be X^R and the gene for anaemia be X^r. Then Susan must be $X^R X^R$ and Alan and Frank must both be normal $X^R Y$ since the father David must be $X^r Y$.

(b) 1 chance in 2 or 50% probability

(c) Frank is $X^R Y$ and Susan must be $X^R X^r$. Thus, the mating is $X^R Y \times X^R X^r$ and the sons must be $X^R Y$ or $X^r Y$. Thus, 50% of the sons must be anaemic.

Questions **5–7** have the answers supplied with the questions.

14 Disease

14.1 Types of disease

Disease is a disordered state of an organ or organism.

1. Diseases caused by other **organisms living parasitically** in the body, e.g. bacteria, viruses, protozoa, worms and fungi.
2. Diseases produced by **aging and degeneration** of body tissues (for diseases of the circulatory system, see page 81). Aging of joints causes arthritic conditions; aging of eye muscles causes long-sightedness, etc.
3. **'Human-induced' diseases** induced in people by themselves either individually or collectively. The group includes domestic and industrial accidents, pollution-related disorders, alcoholism and drug abuse.
4. **Deficiency diseases**, which are still a major problem in developing countries (see Section 4.3).
5. **Genetic and congenital disorders** (inherited diseases and defects present at birth). Children who previously died at infancy are surviving into adulthood and having children of their own.
6. **Mental illness** causes a variety of important disorders. In some countries the majority of hospital beds are occupied by patients suffering from some form of mental illness.

14.2 Transmission of disease

The transmission of living organisms causing disease (see group 1 above) is a major problem in human populations. Owing to their small size, disease organisms can be spread in a variety of ways.

1. **Airborne diseases** (droplet infection). Disease organisms are transmitted in tiny droplets of moisture during exhalation from the lungs, i.e. in breathing, talking, coughing, sneezing, etc. The moisture evaporates, leaving bacteria or virus particles suspended so that they can be inhaled by another person. Droplet-borne infection spreads rapidly during high humidity and crowding, e.g. in schools, buses and trains and at public meetings.
2. **Waterborne diseases**. Drinking-water is a source of many diseases, e.g. dysentery, cholera, typhoid and paratyphoid, which affect the gut. Active pathogens are liberated with faeces and can reach drinking-water where conditions of life are insanitary. The problems are greater when flood, typhoons, earthquakes and so on damage water and sewage systems. Personal hygiene after defaecating or urinating is very important in preventing spread.
3. **Food-borne diseases**. Organisms transmitted by water can also be transmitted by some foods, e.g. bacterial, viral and worm infections. Unwashed hands, septic sores, water and flies can all spread infection to food.
4. **Contagious diseases** are spread by direct contact between people or by objects handled by people, e.g. fungal infections such as ringworm and athlete's foot can be transferred skin to skin or by infected floor coverings and towels.

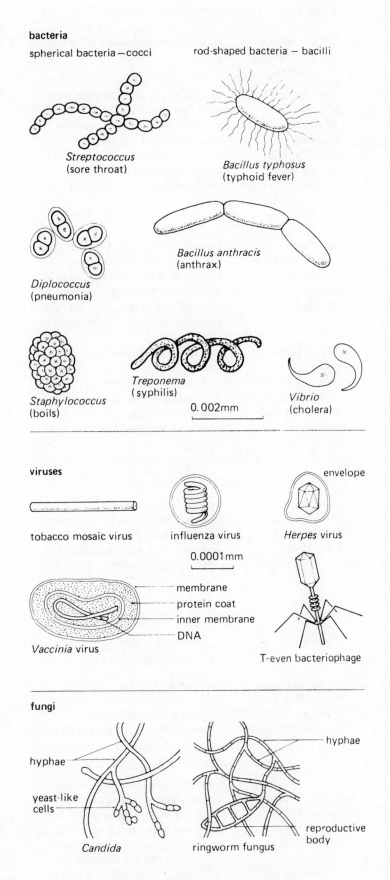

bacteria

spherical bacteria—cocci

rod-shaped bacteria — bacilli

Streptococcus
(sore throat)

Bacillus typhosus
(typhoid fever)

Diplococcus
(pneumonia)

Bacillus anthracis
(anthrax)

Staphylococcus
(boils)

Treponema
(syphilis)

0.002mm

Vibrio
(cholera)

viruses

tobacco mosaic virus

influenza virus

0.0001mm

envelope

Herpes virus

Vaccinia virus

membrane
protein coat
inner membrane
DNA

T-even bacteriophage

fungi

hyphae

yeast-like
cells

Candida

hyphae

ringworm fungus

reproductive
body

Fig. 14.1 Bacteria, viruses and fungi

Table 14.1 Diseases caused by bacteria

Disease, bacterium and method of spread	Symptoms and course of disease	Treatment and control
Tuberculosis (TB) *Mycobacterium tuberculosis* (bacillus) Airborne, occasionally via infected cow's milk	Disease may show itself many years after first infection. Tubercle bacilli may infect many organs, but pulmonary (lung) TB is the most common. General weight loss and cough, sometimes with sputum containing blood; slight afternoon fever.	Treatment: antibiotic drugs, e.g. streptomycin Control: 1. detection by mass radiography 2. vaccination with BCG (Bacille Calmette-Guerin) 3. eradication of cattle TB 4. pasteurisation of milk
Diphtheria *Corynebacterium diphtheriae* (bacillus) Airborne	Incubation period, 2–4 days. Bacteria grow on mucous membranes of respiratory tract, releasing powerful toxin. Slight fever and sore throat are followed by severe damage to heart, nerve cells and adrenal glands.	Treatment: 1. injection of antitoxin 2. antibiotics, e.g. penicillin and erythromycin Control: mass immunisation with diphtheria toxoid
Typhoid and paratyphoid *Salmonella typhi* and *S. paratyphi* (bacilli) Waterborne and foodborne	Incubation period, 6–7 days. Mild fever and slight abdominal pains with constipation, followed by step-like increase in fever level, increasing pain and diarrhoea. Ulceration and rupture of the intestine may occur.	Treatment: antibiotics, e.g. chloramphenicol and ampicillin (semi-synthetic penicillin) Control: 1. purification of water supplies 2. safe disposal of sewage 3. pasteurisation of milk 4. vaccination with killed bacteria
Cholera *Vibrio cholerae* (bacillus) Waterborne	Bacteria release powerful toxin, causing inflammation of the gut and severe diarrhoea ('rice water'). The resulting loss of water and mineral salts may lead to death.	Treatment: 1. injection of saline solution to replace water and salts 2. drugs, e.g. tetracycline, chloramphenicol and sulfadiazine Control: 1. purification of water and treatment of sewage 2. vaccination (3–12 months' protection only)
Whooping cough *Bordetella pertussis* (bacillus) Airborne	Occurs mainly in young children. Incubation period, 2 weeks. Severe coughing bouts, with cough followed by 'whoop' sound as air is inspired through narrowed air-passage.	Treatment: antibiotics Control: vaccination with killed bacteria
Tetanus (lockjaw) *Clostridium tetani* (bacillus) Bacteria present in soil, dust and animal faeces contaminate skin wounds	Toxins produced by bacteria cause muscular spasms (strong contractions), first in the region of the mouth and neck, then throughout the body. Eventually, convulsions may be so severe and frequent that patient dies of exhaustion or lack of oxygen. Commonest in war (infected wounds) and newborn babies (infected stump of umbilical cord)	Treatment: 1. injection of antitoxin from horses 2. injection of muscle-relaxing drugs 3. antibiotics such as penicillin Control: immunisation with tetanus toxoid (toxin which has been treated so that it is no longer harmful)

Table 14.2 Diseases caused by viruses

Disease and method of spread	Symptoms and course of disease	Treatment and control
Measles Airborne	Occurs mainly in young children. Incubation period of 2 weeks, followed by sore throat, runny nose, watery eyes, cough and fever. Small, white spots (Koplik's spots) appear inside mouth on wall of cheek. Two days later, reddish rash appears at hairline, on neck and behind ears, spreading over rest of body. If no complications, patient recovers completely 1 week later, but virus can damage heart muscles, kidneys or brain, and secondary infections by bacteria may cause pneumonia, etc.	Treatment: injection of gamma globulin a few days after exposure Control: 1. live, attenuated vaccine 2. isolation of patient and avoidance of overcrowding in schools
Rubella (German measles) Airborne	Occurs mainly in older children and young adults. Incubation period of $2\frac{1}{2}$ weeks, followed by slight fever and body rash, which disappears after 3 days. Complications rare except in women during first 4 months of pregnancy, when there is a 20% chance of blindness, deafness or other serious defects in the baby.	Treatment: 1. mild analgesics, such as aspirin and paracetamol 2. for pregnant women, injection of gamma globulin within 8 days of exposure Control: vaccination of girls 11–14 years with live, attenuated vaccine
Common cold Airborne	Nasal and bronchial irritation, resulting in sneezing and coughing	Treatment: aspirin, etc. Control: colds can be caused by many different viruses, so vaccine production is impracticable, and people who have recovered from a cold caused by one virus are not immune to colds caused by others
Influenza (sweating sickness) Airborne	Incubation period, 1–3 days. Sudden fever with headache, sore throat and muscular aching. Recovery within 1 week, but after-effects such as tiredness and depression (especially in older people) may last well over a month. Secondary infection of the lung tissue by bacteria, leading to pneumonia, may occur in some cases.	Treatment: 1. analgesics 2. antibiotics to prevent secondary infections Control: vaccination with in-activated virus may give immunity for 1–2 years
Poliomyelitis (infantile paralysis) Airborne, food-borne or waterborne, also contagious	Incubation period, 7–12 days, followed by fever, headache and feeling of stiffness in neck and other muscles. Virus destroys nerve cells that supply muscles, causing paralysis and muscle-wasting. If breathing muscles are paralysed, an 'iron lung' may be needed. Many people have a very mild form of the disease which they do not notice but which makes them immune. Most cases of paralysis occur in children 4–12 years, but adults may also be affected	Treatment: no known drug or chemical treatment Control: vaccination with formaldehyde-treated virus (Salk vaccine) or mutated virus (Sabin vaccine) taken orally

Table 14.3 Protozoan diseases of Man

Disease	Cause	Transmission	Symptoms, other characteristics and treatment
Malaria	*Plasmodium* spp.	*Anopheles* mosquito bite	Plasmodia injected into the blood multiply rapidly. After 10 days, high fever develops which may be continuous, irregular or occur twice a day. Control: 1. drainage of the breeding places of mosquitoes 2. destruction of larvae with an oil spray 3. destruction of adults with insecticide 4. destruction of the parasites in Man by drugs, e.g. chloroquine and quinine 5. preventive drugs, e.g. proguanil and pyrimethamine
Amoebiasis (amoebic dysentery)	*Entamoeba histolytica*	Uncooked food; un-hygienic preparation of food	Causes diarrhoea with loss of blood, fever, nausea and vomiting – can lead to death. Control: 1. hygienic food handling 2. prevention of flies that can spread the disease 3. drugs, e.g. emetine, antibiotics and sulphur drugs

Table 14.4 Fungal diseases of Man

Disease	Cause	Symptoms, other characteristics and treatment
Ringworm of the scalp	*Microsporium audouini*	A highly contagious disease by contact, combs, hats, etc., among children. It begins as a small scaly spot which enlarges, and older patches are covered with greyish scales. Control: 1. exclusion of infected children from school 2. drugs, e.g. antibiotic griseofulvin, taken by mouth
Ringworm of the skin	As for scalp, or *M. canis*	Lesions on the skin are seen as pale, scaly discs. There is more inflammation around the edges, causing swelling and blistering. Control: drugs, e.g. griseofulvin
Athlete's foot	*Tinea pedis*	Shows as sodden, peeling skin between the toes which can be subject to secondary bacterial infection. Cure rate is low. Control: 1. exclude sufferers from swimming pools and changing rooms 2. griseofulvin is only used in extreme cases
Candidiasis (thrush)	*Candida albicans*	A yeast-like cell, 2–4 μm in diameter. Commonly harmless in the body, but infection results from some local reduction in resistance of the tissues. This may occur in the mouth, intestine, vagina, etc. Control: 1. establish the predisposing factor and change this to clear up the infection 2. drug, e.g. antibiotic nystatin, used as a local cream

Table 14.5 Worms infecting Man

Disease and worm	Transmission	Hosts	Symptoms, other characteristics and control
Tapeworm *Taenia* spp.	Through food – undercooked meat and fish	1. Man 2. Pig, cattle or fish	The encysted embryo in the flesh of secondary host is consumed in undercooked meat or fish. Tapeworm develops in the gut attached to the intestinal wall. Fertilised eggs are passed out with the faeces; eggs are then eaten by animals. The tapeworms cause few symptoms and relatively little damage in Man. Control: 1. meat and fish to be well cooked at high temperatures 2. inspection of meat at slaughterhouses 3. proper processing and disposal of sewage 4. drugs, e.g. similar drugs to those used against malaria: mepacrine and chloroquine
Ascariasis *Ascaris lumbricoides*	Infected food and water	Man (no secondary host)	The worms live in the bowel of Man and produce vast numbers of eggs that are very resistant when shed with the faeces. When eaten by Man, the eggs hatch and the larvae burrow into the lungs and from there reach the gut by way of the pharynx. The worms may obstruct the bowel and the larvae damage the lungs, causing malnutrition and death. Control: 1. proper processing and treatment of sewage 2. hygienic food and water supply 3. drugs; piperazine is the most effective
Threadworms or pinworms *Enterobius vermicularis*	Eggs swallowed	Man (no secondary host)	Very common, especially in children. Adults live in the large intestine. Females migrate to the anus to lay eggs, causing itching. Scratching, followed by placing of fingers in mouth, causes reinfection. Control: washing hands after using toilet or touching anal area

5. **Sexually transmitted diseases (STDs)**. Venereal diseases are spread by intimate sexual contact. The most widely known are syphilis and gonorrhoea. Herpes is a more recent addition to the list of STDs.
6. **Insect-borne diseases**. Many kinds of microbes are spread by insects (**vectors**), e.g. houseflies carry intestinal disease-causing organisms on their body and through their gut; mosquitoes carry malaria parasites and yellow fever virus; fleas carry bacteria causing bubonic plague.

14.3 Cancer

The term 'cancer' describes the condition in which certain cells continue to divide in spite of close contact with surrounding cells and form a **tumour**. There are two types: (1) **benign** (harmless) and (2) **malignant** (harmful).

Malignant tumours are capable of invading surrounding tissues, and also may be spread by means of the blood and lymphatic systems to other parts of the body,

Table 14.6 Sexually transmitted diseases (STDs)

Disease and pathogen	Incubation period	Symptoms	Control
Gonorrhoea A bacterium: *Neisseria gonorrhoeae*	2–6 days	Men: inflammation and discharge from the penis; pain on urinating Women: inflammation of the vagina and urethritis; may be present without noticeable symptoms	Antibiotics such as penicillin and sulphonamide drugs
Syphilis A tryponeme: *Tryponema pallidum*	2–4 weeks	Ulceration at the point of entry of the tryponeme. Usually on sex organs but may be on lips, tongue or anus	Antibiotics such as penicillin in the early stages
	3–14 weeks	Secondary syphilis: develops fever, sore throat, swollen lymph glands up to 2 years after first infection, then symptoms disappear	
	after 5 years	Tertiary syphilis: non-infectious; can result in insanity and heart problems	
Genital herpes A virus: herpesvirus type II	2–7 days	Small ulcers on the genital organs	

where **secondary tumours** are established. This disease is one of the major causes of death in the developed world. It is usually a disease of old age. Treatment is by (a) **surgery**, (b) **radiotherapy** (killing cells by ionising radiations) and (c) **chemotherapy** with drugs. All of these treatments may be used on the same patient.

14.4 Mental health

A high incidence of mental problems is found in the populations of the countries of the developed world. Some 50% of all hospital beds are occupied by people suffering from some form of psychiatric disorder.

Classification of mental disorder is difficult, but it can be considered under three main headings:

1. **Congenital handicaps**, e.g. mongolism, brain damage during birth.
2. **Senility** — feeble-mindedness of old age.
3. **Control disorders** — neuroses and psychoses.

    (a) **Neuroses** include depression, overanxiety, claustrophobia (fear of enclosed spaces), agorophobia (fear of open spaces) and compulsive cleanliness.
    (b) **Psychoses** usually represent more serious lack of control and are correspondingly more difficult to treat. They include schizophrenia, which accounts for about half of all cases of severe mental illness. The patient tends to lose contact with reality and suffers from strong hallucinations. Psychopathic personality also comes under this heading. The condition is characterised by behaviour entirely governed by impulse. This leads to a great deal of criminality and anti-social behaviour without any trace of remorse.

Mental health is difficult to define, but it is generally agreed that **a mentally healthy person should have sufficient intelligence and control over his/her emo-**

tions to care adequately for himself/herself and his/her family, and to make a positive contribution to society. The factors contributing to this state are (a) **physical well-being**, (2) **emotional stability** and (3) **high self-esteem**. The factors contributing to mental illness are (1) **inability to cope with stress**, (2) **low self-esteem** and (3) **past experience of problems**.

14.5 Questions and answers

(a) Multiple-choice Questions

Questions **1–6** refer to the scientists listed below.

A Leeuwenhoek
B Pasteur
C Koch
D Spallanzani
E Lister

1 Which one of the above first demonstrated that decay is caused by living organisms?
2 Which one of the above first cultured the bacteria that cause anthrax?
3 Which one of the above is credited with inventing the light microscope?
4 Which one of the above first developed the germ theory of disease?
5 Which one of the above first used carbolic acid as a disinfectant?
6 Which one of the above developed a method of culturing micro-organisms by putting them on a sterile jelly?

Questions **7–11** refer to the organisms listed below.

A staphylococci
B bacilli
C vibrios
D spirochaetes
E streptococci

7 Which one of the above causes cholera?
8 Which one of the above causes syphilis?
9 Which one of the above causes boils and pimples?
10 Which one of the above has a rod-shaped structure?
11 Which one of the above causes a sore throat?

12 Viruses consist of
  A nucleic acid and protein.
  B protein and lipid.
  C lipid and carbohydrate.
  D carbohydrate and nucleic acid.
  E protein and carbohydrate.

13 Which one of the following diseases is caused by a protozoon?
  A chicken pox
  B malaria
  C pneumonia
  D influenza
  E measles

14 Athlete's foot is caused by a
  A bacterium
  B virus
  C fungus
  D protozoon
  E flatworm

Select the appropriate letter **A–E** below to indicate the method of transmission of each of the diseases listed in questions **15–22**.

A transmission by air
B transmission by contact
C transmission by insects
D transmission by sexual contact
E transmission by water

15 Cholera
16 Influenza
17 Gonorrhoea
18 Ringworm
19 Malaria
20 Typhoid
21 Herpes genitalis
22 German measles

23 Which of the following defines a pathogen?
   A bacteria and fungi
   B poisonous chemicals
   C symbiotic bacteria
   D organisms that cause disease
   E parasitic tapeworms

24 Which of the following is the method by which a virus multiplies?
   A binary fission
   B asexual reproduction
   C spore production
   D using the metabolism of the host cell
   E using mitosis in the host cell

25 Which one of the following describes a disease which spreads world-wide?
   A sporadic
   B virulent
   C epidemic
   D pandemic
   E endemic

26 Which of the following methods should be used to control the spread of cholera?
   A inoculation of cholera victims
   B avoiding wading in water
   C chlorinating drinking-water
   D spraying oil on water
   E adding salt to water

27 Which one of the following diseases is caused by a spirochaete?
   A influenza
   B sleeping sickness
   C tuberculosis
   D syphilis
   E typhus

28 In which one of the following could a virus develop?
   A milk
   B agar jelly
   C protein broth
   D chick embryo
   E moist bread

(b) Structured Questions

1 (a) Measles is by far the most prevalent infectious childhood disease in the United Kingdom. It is spread by droplet infection and its complications are among the most serious. A small percentage of children suffer a potentially serious complication,

middle ear infection, which may lead to deafness. Immunisation is encouraged, widely available and free. However, only 58% of the children in the United Kingdom are now vaccinated, even though a vaccine has been available for many years.

(i) In what specific ways does droplet infection cause the spread of disease?

(ii) Why should an infection of the middle ear lead to deafness when the sensory cells are located in the inner ear?

(iii) What is a vaccine?

(iv) How does vaccination prevent the occurrence of measles?

(v) Identify **two** possible causes for the low percentage of children vaccinated against measles.

(vi) It has been estimated that if the vaccination rate for measles rose to 83% the incidence of the disease would fall to almost zero. Explain why a 100% vaccination rate is not necessary to achieve this result. **(12)**

(b) Many insects transmit organisms without showing any symptoms of disease themselves. They are said to be vectors of the disease which is caused by the pathogens they transmit.

For any **named** human disease transmitted by insects identify

(i) the type of agent causing the disease

(ii) the vector

(iii) how the agent is transferred from the vector to man

(iv) **two** methods of controlling the vector. **(5)**

(c) 'The enjoyment of the highest attainable standard of health is one of the fundamental rights of every human being' is an extract from a statement on health from the World Health Organization. Indicate clearly **two** ways in which the Civil Authorities at seaports or airports attempt to achieve the above objective for the resident population of the country. **(4)**

(NISEC)

2 The graph shows the North American death rate from typhoid (a water-borne disease) between 1904 and 1918. Chlorination of the domestic water supplies was introduced in 1908 to destroy disease organisms.

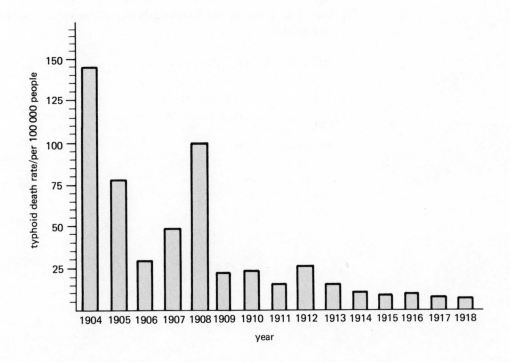

(a) (i) What was the death rate per 100 000 people from typhoid in

1906 . . . .

1907 . . . .

1908? . . . . **(3)**

      (ii) Suggest **one** reason why there was a variation in the death rate over these three years. (two lines) **(1)**

(b) (i) What can be concluded about the death rate from typhoid from 1909 onwards compared with the death rate from 1904–1908? (two lines) **(1)**

      (ii) Explain why there was a difference in overall death rate between these two periods of time? (three lines) **(1)**

(c) State **two** ways, other than by water, that infectious disease can be transmitted. (two lines) **(2)**

(d) Typhoid is a disease caused by a bacterium.

     (i) Name another type of organism which is the cause of an infectious disease. (one line) **(1)**

     (ii) Name a disease caused by this organism. (one line) **(1)**

(UCLES)

3 (a) In the figure below, indicate the four stages, in their correct order, in the life cycle of the common housefly. **(4)**

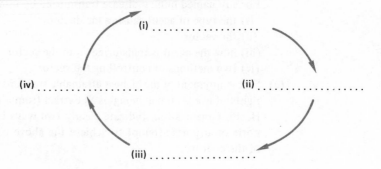

(b) Name three diseases spread by houseflies.

   (i) . . . . (ii) . . . .

   (iii) . . . . **(3)**

(c) Give three precautions a houswife should take to reduce contamination of foodstuffs by houseflies.

   (i) . . . .

   (ii) . . . .

   (iii) . . . . **(3)**

(OLE)

4 Methods of controlling the spread of disease can include
  destroying the breeding grounds of the vector – A
  separating the infected individuals from the general public – B

For each of the diseases listed in the table below, in the right hand column, write
A – if this is more effective
B – if this is more effective
AB – if both are effective
N – if neither is effective

Disease	Method of control
Influenza	
Cancer	
Syphilis	
Malaria	
Scurvy	
Diphtheria	
Typhoid	
Tuberculosis	

(8)
(L)

5 Examine the drawings of bacteria in the figure below.

(i) . . . . (ii) . . . . (iii) . . . . (iv) . . . .

(a) In the spaces (i) to (iv) state the type of bacteria that each drawing represents. **(4)**
(b) State **three** examples of bacteria that are harmful to Man and in each case state the name of the disease that they produce.
   Bacterium . . . . Name of disease . . . . **(6)**
(c) Some disease-causing organisms (pathogens) are carried by other organisms (vectors). State **two** examples of a vector and the disease that it may transmit to humans.
   Vector . . . . Disease transmitted . . . . **(4)**
(d) What is meant by natural immunity? (two lines) **(2)**

6 The diagram below shows bacteria seen through a microscope. Study it carefully and answer the questions below.

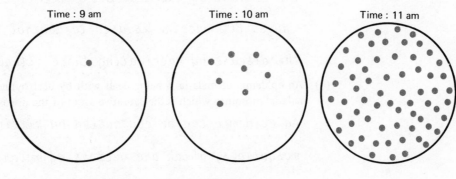

Time : 9 am   Time : 10 am   Time : 11 am

(a) How many bacteria are present after 1 hour? . . . . 8 . . . . **(1)**
(b) How many bacteria are present after 2 hours? . . . . 64 . . . . **(1)**

253

(c) Bacteria divide once every 20 minutes. How many bacteria would be present after 1 hour 20 minutes? . . . . 8 . . . . **(1)**

(d) Name **two** conditions which are necessary for bacteria to increase rapidly in number.

1. *Optimum temperature* . . . .

2. *Adequate food supply* . . . . **(2)**

(EMREB)

7 The following table has columns for diseases, point or method of entry and causative agents. Complete the table by filling in the spaces provided. Select your answers **only** from the lists below.

Diseases: bilharziasis, dengue, filariasis, malaria, plague, syphilis, typhoid.

Point or method of entry: insect bite, intestine, nose, sexual organs, skin, vagina, wound.

Causative agents: bacteria, fluke, fungus, protozoa, spirochaetes, virus, worm.

Disease	Point or method of entry	Causative agent
Hookworm	*Skin*	Worm
Typhoid	Intestine	Bacteria
Syphilis	Sexual organs	*Spirochaetes*
Influenza	*Nose*	*Virus*
Malaria	*Insect bite*	Protozoa
Dengue	*Insect bite*	*Virus*
Tetanus	*Wound*	*Bacteria*

(6)

(UCLES)

8 In each of the following news items a biological error has been made. Explain what is wrong, and why.

(a) 'It is reported that one of the England football team has a cold and is being treated with antibiotics.'

A cold is caused by a virus infection.

Viruses cannot be treated by antibiotics and

therefore the statement is incorrect. **(2)**

(b) 'Outbreaks of tuberculosis have occurred in some inner city areas. Patients are being isolated and treated with BCG vaccine.'

The control of tuberculosis patients is by means of

drugs. BCG vaccine is used to control the spread of

the disease by vaccinating those without immunity. **(2)**

(c) 'An epidemic of malaria is being dealt with by destroying the breeding places of the malarial mosquito which is the causative agent of the disease.'

An epidemic can be controlled by killing off

mosquitoes, which are vectors of malaria. They are not

the causative agent - this is a protozoon, Plasmodium. **(2)**

(L)

9 This question is about diseases and their control.
   (a) Organisms that cause disease are parasites.
       (i) Explain what is meant by the term parasite. . . . .

       . . . . **(1)**
       (ii) Name **two** groups of organisms that cause human disease.

       (1) . . . . (2) . . . .
                                                                                         (2)
       (iii) State **two** ways that disease organisms can enter the human body.

       (1) . . . . (2) . . . .
                                                                                         (2)
   (b) Complete the following sentences by writing in the spaces words that you think
       make the best sense.
       'After recovering from a disease such as mumps or chicken-pox, a person is usually

       . . . . to further attacks of the disease because the white cells of the blood have

       produced . . . . For other diseases such as whooping-cough and measles these can
       be formed by having a . . . .' **(3)**

   (c) A plate of nutrient agar had three antibiotic discs placed on it. A bacterial sample
       was added, and the plate incubated for 48 hours at body temperature. The results are
       shown below.

(Dots represent bacterial colonies)

       (i) Which was the most effective antibiotic against the bacterial culture?

       . . . .
       (ii) Which antibiotic did not affect the bacterial culture?

       . . . .
       (iii) Which would be the best antibiotic for a doctor to prescribe to a patient having a
       disease caused by these bacteria?

       . . . .
       (iv) Name one antibiotic.

       . . . .
       (v) A doctor prescribed a course of antibiotics to be taken for five days. A patient
       felt better after two days and stopped taking his antibiotics. A week later the
       disease occurred again. Suggest a reason why this happened.

       . . . .

       . . . . **(5)**

(d) (i) Gonorrhoea is a type of sexually transmitted disease. How is it transmitted from one person to another?

. . . .

(ii) What type of micro-organism causes gonorrhoea?

. . . .

(iii) Give **two** possible symptoms of gonorrhoea in the male.

(1) . . . .

(2) . . . .

(iv) What is the modern treatment for gonorrhoea likely to be?

. . . . **(5)**

(e) Humans suffer from diseases not caused by parasites. State an example of:

(i) a sex-linked inherited disease . . . .

(ii) a lung disease . . . . **(2)**

(LREB)

Free-response Question

(a) What is meant by the statement 'Tuberculosis is endemic in most countries of the world'?
(2)

(b) What is meant by an epidemic? **(2)**

(c) What are the signs and symptoms of pulmonary tuberculosis? **(7)**

(d) How can the spread of tuberculosis be controlled? **(9)**

(UCLES)

Answer

(a) The term 'endemic' in this sentence means that some cases of the disease are always present but that the numbers are usually low. This applies to most countries of the world.

(b) The term 'epidemic' referring to a disease means that an outbreak of the disease develops rapidly. The number of cases of the disease increases dramatically, resulting in a large proportion of the population being affected.

(c) The disease of pulmonary tuberculosis affects the lungs and the breathing of those suffering from the infection. There is a persistent cough, loss of weight, tiredness and fatigue, and the individual looks pale. There are intermittent bouts of fever. As the disease progresses, the patient develops lesions on the lungs and these result in blood and mucus being coughed up and spat out. An X-ray examination of the lungs will show infection and lesions as shadows on the pale background of the lungs in the X-ray pictures.

(d) The spreading of the disease of tuberculosis can be controlled first by the obvious hygienic methods to prevent the bacteria reaching the air. The mouth should always be covered when coughing. Spitting should be discouraged, but if it cannot be avoided, then lidded containers must be provided. Ventilation of rooms and meeting places is essential to decrease humidity and thus provide conditions where the bacteria cannot flourish.

To avoid the disease spreading in a population, the use of mass X-rays can provide early detection of the disease and then proper treatment. Young children should be given immunity by vaccination with BCG. This can also be given to any members of a family where there is an infected individual. Patients with advanced stages of the disease should be isolated.

Pulmonary tuberculosis bacilli can be carried in cattle and transmitted to humans through the marketing of the milk. Most milk comes from cattle which have been tuberculin tested (TT), indicating that the herd has been treated to ensure that it is free from the disease. Highly infected cattle should be killed. A further precaution is that all milk should be pasteurised, a process that ensures the death of all pathogenic bacteria.

Other hygienic measures such as the sterilisation of all containers used by patients and the control of houseflies will all contribute towards containing and preventing the spread of the disease.

(c) Answers to Objective and Structured Questions

(i) *Multiple-choice Questions*

1. D   2. C   3. A   4. B   5. E   6. B   7. C   8. D   9. A   10. B   11. E   12. A
13. B   14. C   15. E   16. A   17. D   18. B   19. C   20. E   21. D   22. A
23. D   24. D   25. D   26. C   27. D   28. D

(ii) *Structured Questions*

1 (a) (i) Coughing, sneezing and shouting all expel droplets of moisture which could contain virus particles. When the moisture droplet dries up, the virus is left floating in the air and could then be inhaled.

    (ii) The ear ossicles transmit sound from the eardrum to the inner ear and are essential for hearing. Infection of the middle ear can damage or inhibit the work of these ossicles and so lead to deafness.

    (iii) It is a toxoid, killed bacteria or weak strain of live bacteria or virus.

    (iv) It induces the body to produce antibodies in response to the antigens of the contents of the vaccine.

    (v) Parents are afraid of the complications which might arise from vaccination, e.g. brain damage.

      Parents are willing to take the risk that their child has natural immunity.

    (vi) It would appear that in a large population about 17% of children would be naturally immune or immune as a result of a mild dose of the disease.

(b) There are a number of examples that can be given here. The most common example to be used in examinations is malaria.

Malaria (i) Protozoon (*Plasmodium*)
        (ii) Mosquito   (ii) Insect bite   (iv) Spraying the breeding place with oil to kill larvae and pupae *or* using insecticides to kill adult mosquitoes in houses.

(c) Spray aircraft entering the country. The authorities should use insecticide to kill insect vectors and spray all internal spaces of the plane.

    Ensure that immigrants are given a thorough medical check on arrival, to prevent the entry of tropical disease and other diseases that have been eradicated in the country of arrival.

2 (a) (i) 1906, 30; 1907, 50; 1908, 100

    (ii) Rapid increase in populations (immigration) into the large cities without the health facilities to accommodate them.

(b) (i) There was a rapid decrease between 1908 and 1909 down to 25 per 100 000. This level was then maintained and even decreased from 1909 to 1918.

  (ii) 1906–1908 the water supply was infected with cholera vibrios. After that year chlorination destroyed these vibrios and ensured a clean water supply.

(c) Insects and air (d) (i) Virus (ii) Influenza

3 (a) (i) Imago/adult (ii) Egg (iii) Larva (iv) Pupa

(b) Dysentery; food poisoning; cholera

(c) 1. Cover all food with muslin or wrap in polythene.

  2. Keep food in a refrigerator or meat safe.

  3. Suspend insecticide containers in the kichen.

4 The table should be completed as follows:
influenza, B; cancer, N; syphilis, B; malaria, AB; scurvy, N; diphtheria, B; typhoid, B; tuberculosis, B

5 (a) (i) Streptococci (ii) Bacilli (iii) Vibrios (iv) Spirochaetes

(b) *Clostridium*, lockjaw; *Vibrio*, cholera; *Salmonella*, food poisoning;

(c) Mosquito, malaria; tsetse fly, sleeping sickness

(d) Inherited immunity or immunity developed as a result of suffering from a disease.

Questions **6–8** have the answers supplied with the questions. Question **9** has no answers supplied. Try completing this question yourself.

15 Health and Hygiene

15.1 Treatment of disease

1. **Natural protection** of the body (see Chapter 14 and free-response question 1 in this chapter).
2. **Immunity and vaccination** – this method is used to supplement the body's natural defences. See Table 15.1 for a summary of types of immunity and Table 15.2 for an immunisation schedule for children.

Table 15.1   Punnett square, showing types of immunity

	Active (involves production of antibodies by patient's lymphocytes)	*Passive* (involves use of antibodies produced by another animal)
Natural	(a) inherited (b) acquired on recovery from disease	antibodies received via placenta or mammary glands
Artificial	from vaccination	from injection of serum containing antibodies

3. **Treatment of patient disease**. Bacterial disease can be treated by:
   (a) Neutralising bacterial toxins, which is achieved by inoculation with specific antitoxins obtained from animals, e.g. diphtheria antitoxin.
   (b) Preventing bacteria from multiplying.
   (c) Killing bacteria.

   (b) and (c) involve the use of drugs.

Table 15.2   Typical immunisation schedule for children

Disease	*Type of vaccine*	*Age for immunisation*	*Age for booster doses*
Diphtheria	Toxoid	Given in	
Tetanus	Toxoid	three doses	5 years ⎱ 17 years
Whooping cough	Killed bacteria	at 4, 6 and	
Poliomyelitis	Attenuated virus (given by mouth)	12 months	17 years
Measles	Attenuated virus	18 months	None – lifelong immunity
German measles (rubella)	Attenuated virus	13 years (girls only)	None – lifelong immunity
Tuberculosis	Attenuated bacteria	14 years (after testing to show no natural immunity)	None – lifelong immunity

Table 15.3 Types of vaccine

| Vaccine production | Antibodies produced against diseases | |
	Bacterial disease	Viral disease
Living attenuated micro-organisms	Tuberculosis	Yellow fever, measles, german measles, poliomyelitis, rabies
Dead micro-organisms	Typhoid, paratyphoid, whooping cough, cholera	Influenza
Toxoids	Tetanus, diphtheria	

(i) **Chemotherapy**. Over the centuries many drugs have been obtained from naturally occurring substances. Drugs have also been chemically synthesised which prevent bacteria growing, e.g. sulphonamides. Since they prevent growth, they are called bacteriostatic drugs. Use of these has declined in favour of bactericidal drugs, which act more quickly by killing bacteria.

(ii) **Antibiotics** are a major group of drugs derived from extracts of bacteria and fungi. The first antibiotic was discovered in 1928 by Fleming, who extracted penicillin from a mould, *Penicillium*. Antibiotics became generally available in the 1940s. Penicillin has two major advantages over other drugs: it is harmless to humans, even in large doses, and it is effective against a wide range of diseases.

Other antibiotics such as tetracyclines and chloramphenicol may be used against some of the larger viruses. Griseofulvin is an antibiotic that acts against fungi. In general, antibiotics do not act against viruses.

15.2 Drug abuse

Drugs used for treating disease can be misused by people. Unfortunately such mis-use can lead to **addiction**, which means that the user is unable to do without a particular drug. Addiction develops through the following stages:

1. **pleasant feelings** during first use, leading to
2. **repetition**, and then to
3. **tolerance**. The body now needs more and more of the drug to satisfy. Any attempts to give up lead to
4. **withdrawal**. This condition shows symptoms caused by absence of the drug in the body and the craving for its renewal.

(a) Classification of Drugs

Drugs used by Man can be classified as follows.

(i) *Domestic Drugs*

1. **Caffeine** in tea or coffee – not usually addictive.
2. **Alcohol** – the only true drug producing marked psychological effects which is socially acceptable in the West. Only one in 500 partakers will become an alcoholic. There are about 100 000 acute alcoholics in England and Wales.

Alcohol is not a stimulant but a sedative, and in large doses reduces the drinker to unconsciousness. It acts in the same way as anaesthetics such as ether or chloroform. Used in small amounts, it can remove inhibitions and produce talkativeness (it acts to break the ice at parties). Increased intake, however, leads to an inability to concentrate and forgetfulness, and leads eventually to slurred speech, inability to walk and finally collapse.

Cure of alcoholism is not possible, as an alcoholic can never drink safely. The condition can be treated when all alcohol is denied and the alcohol in the tissues gradually disappears.

3. **Nicotine** is present in tobacco, which can be smoked in various ways. In the UK cigarette smoking is responsible for 90% of lung cancer deaths, 75% of deaths from bronchitis and 25% of deaths from coronary heart disease. The effects of smoking have been shown by many studies to defeat the advances made by medicine to increase the life-span.

(ii) *Hallucinogenic Drugs*

Cannabis (marijuana, hashish, pot, grass) is obtained from the leaves and flowers of the hemp plant. Its use is widespread throughout the world. It is smoked in the form of a hand-made cigarette. Its effect on the body is to produce a pleasant happy feeling. It is probably not addictive but can lead to the use of hard drugs such as heroin. Its use is illegal in most countries.

(iii) *Barbiturates*

Barbiturates are depressive drugs. They include sodium amytal, and phenobarbitone. They are used generally as sleeping tablets but, like other drugs, their use can be abused.

(iv) *Stimulants*

Stimulants include the so-called 'pep' pills, which produce a feeling of well-being and wakefulness. They are very widely used and tolerance develops quickly.

(v) *Opiates*

Opium is the product of the opium poppy (*Papaver somniferum*). It contains the pain-killing chemicals **morphine** and **codeine**. The drug **heroin** is derived from morphine and at the present time is being used in epidemic proportions throughout the world. Heroin is responsible for addiction and death. The traffic in the drug involves millions of pounds each week.

'Withdrawal sickness' from heroin or the other opiates is a shattering experience for the addicts, since it involves fever, shivering, aching bones, nausea, vomiting and diarrhoea.

(vi) *Solvents*

The sniffing of solvents is a comparatively recent phenomenon. It involves the inhalation of volatile solvents of glues and resins. Shopkeepers have to consider very carefully whether they are selling this material for genuine use or not.

15.3 Personal hygiene

(a) Skin

1. The skin is continually in contact with micro-organisms. Those on the fingers and hands can make contact with food and the mouth. All body skin must be kept clean but, in particular, the hands. Soap and water are effective in keeping the skin clean.
2. The hands should be washed after using the lavatory, to prevent the spread of intestinal diseases.
3. The fingernails always carry bacteria, and therefore they should not be used to pick at spots or wound scabs.
4. Feet must be kept clean, particularly between the toes. Wear properly ventilated shoes and socks. Bare feet do not usually pick up pathogens, but the fungal infection athlete's foot can be picked up from wet floors, wooden drainage boards and towels. Regular washing and thorough drying should prevent this disease.
5. Skin parasites such as the itch mite causing scabies, lice and fleas can be avoided by regular changes and washing of underclothing. In addition, bathing with soap and water together with vigorous scrubbing with a nail brush is essential.

(b) Hair

The hair can be infected with parasites. It must be kept clean by brushing and combing every day and washing at least once per week. Brushing improves the blood circulation in the scalp and washing removes natural grease which collects dirt.

Dandruff can be treated by shampoos containing selenium compounds. Infection by head lice has increased in recent years. They can cause intense itching and their presence is recognised by the discovery of nits on the hairs or even adult lice. Dusting powders, creams and shampoos are available containing gamma-BHC (benzene hexachloride).

(c) Teeth

See Chapter 5.

(d) Nose

The hairs in the nostrils act as a filter against dust and micro-organisms. Breathing should be through the nose and not by way of the mouth and buccal cavity. The nose should be blown using cotton or paper disposable handkerchiefs, in order to clear the nasal passages (see Section 14.2).

15.4 Control of vectors

(a) Houseflies

See Fig. 15.1 for the life cycle. Houseflies carry pathogens from dung: (1) on the legs (bristles), (2) on the body (bristles), (3) on the proboscis, through which they

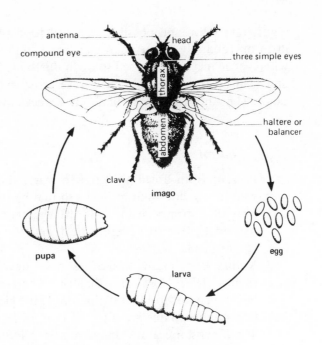

Fig. 15.1   Life cycle of the housefly

Table 15.4   Food poisoning

Causative agent	Source	Symptoms	Prevention of transmission
Salmonella spp., causing salmonelliasis	Many animals carry the disease organisms, e.g. pigs, calves and poultry. Main source for Man is meat of these animals	Within 12–24 hours, fever followed by vomiting and diarrhoea. Firm diagnosis needs laboratory tests on faeces. Rarely fatal	Flies can transmit the bacteria which are excreted in the faeces; therefore environmental control is important. This is difficult to achieve on farms. Fish and shellfish can transmit salmonella, especially if in contact with sewage. Refrigeration, complete thawing and thorough cooking will contain and eventually kill bacteria
Clostridium welchii, causing clostridial food poisoning	Widely distributed in nature, e.g. soil, sewage and water. Thus, many possibilities of food contamination. Spores can survive several hours of boiling water. Many outbreaks traced to meat	12–24 hours incubation; followed by fever and vomiting, abdominal pain and diarrhoea. Infection only lasts about 24 hours; rarely fatal	Meat should be thoroughly roasted in small quantities, not more than 2–3 kg. Meat should be eaten after cooking, any remaining refrigerated within 1½ hours
Clostridium botulinum, causing botulism	Anaerobic bacteria living where air is excluded, as in canned, potted or pickled food. The food generally has been treated but spore forms survive	Rare disease, but a high mortality rate – 50% in reported cases. Takes 24 hours to develop, with vomiting, muscle paralysis and constipation	Adequate heating of food will destroy the spores. Growth of the organism can be inhibited by complete drying, refrigeration, thorough salting or reduction of pH, i.e. acid conditions

regurgitate saliva and possibly digested food (dung), (4) in the faeces, which they drop on food (fly dirt). Their eggs are laid in faeces, manure and rotting material wherever it may be exposed in open dustbins, farmyards, and so on. Larvae hatch and feed on this material. The larvae grow, moult twice and then pupate in a dry place. The adult flies hatch in about 4 weeks in summer.

(i) *Control*

1. **Prevent from breeding**. Kitchen waste should be placed in a dustbin with a well-fitting lid. Manure heaps should be turned regularly, so that the internal heat of decomposition kills the eggs and larvae. Compost heaps should be packed tightly to generate heat and the surface should be covered with a layer of soil.
2. **Killing flies**. Many houseflies have developed resistance to various sprays such as DDT, BHC and dieldrin. There are, however, effective compounds based on pyrethroids and piperonyl butoxide which can be included in sprays. Organophosphates are also used on cards and strips to be hung from ceilings. These substances are dangerous to handle and the vapour is poisonous in small confined spaces.
3. **Protecting food from houseflies**. Shopkeepers must keep displayed food protected in cold cabinets, behind gauze or plastics, to prevent flies settling. Meat and wet fish are particularly attractive to flies, and when brought into the home should immediately be put away into a refrigerator, ventilated cupboard or meat safe. If left on a kitchen bench, the food should be covered with gauze (to allow air circulation but keep away flies).

(b) Mosquitoes

See Fig. 15.2 for the life cycle. Mosquitoes carry the malaria-causing parasite (**Plasmodium**). There are several species of *Plasmodium*, each causing fevers of different duration. The parasite lives in the red blood cells and liver cells of Man.

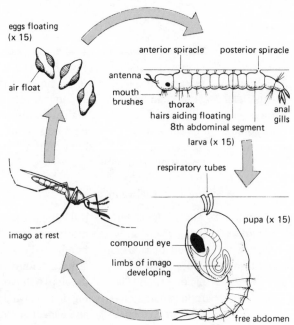

Fig. 15.2   Life cycle of the *Anopheles* mosquito

A stage in the life history of *Plasmodium*, called the sporozoite, develops in the salivary gland of the female mosquito. The insect when feeding thrusts the feeding tube (stylets) through the human's skin and injects saliva. This fluid contains a substance to prevent blood clotting and at the same time hundreds of malarial parasites are introduced into the bloodstream.

(i) *Control*

A combination of methods is necessary for successful elimination of the vector and thus the disease of malaria.

1. **Prevent from breeding**. Lakes, ponds and any still water (swamps, roof gutters, drains) should be sprayed with light oil and insecticide. The oil, remaining on the water surface, prevents larvae and pupae from taking air. The introduction of fish that feed on mosquito larvae helps to reduce the number of vectors. Swampy ground should be drained.
2. **Killing insects**. Insecticides, first introduced in 1935, have had considerable success in reduction of the world-wide incidence of mosquitoes and malaria. The insecticide is sprayed inside buildings, particularly near ceilings and under roof eaves, where mosquitoes rest. Insecticides used are DDT, gamma-BHC and dieldrin. A problem has been the persistence of these insecticides and their concentration in natural food chains (see Fig. 16.2).
3. **Personal protection**. Windows and doors of houses can be protected by fly gauze to prevent entry of mosquitoes at dusk. Humans can further protect themselves by wearing long sleeves and trousers to prevent the bite of mosquitoes. To prevent biting during sleep, the bed should be covered with a mosquito net.

15.5 Questions and answers

(a) Multiple-choice Questions

1 Which one of the following organisms is responsible for food poisoning in humans?
  A virus
  B protozoan
  C *Salmonella*
  D *Clostridium*
  E *Plasmodium*
2 The treatment of disease by the use of drugs is called
  A vaccination.
  B immunisation.
  C inoculation.
  D chemotherapy.
  E antisepsis.
3 Which of the following produces the antibiotic penicillin?
  A a virus
  B a bacillus
  C a bacterium
  D a fungus
  E a parasite

4 In which of the following do *both* drugs change the metabolism of a person and thus lead to addiction?

   A caffeine and morphine

   B aspirin and cannabis

   C paracetamol and heroin

   D morphine and heroin

   E cannabis and caffeine

5 Which one of the following is a dangerous, immediate reaction of the body as a result of taking alcoholic drinks?

   A Blood vessels in the skin contract.

   B Liver cells break down.

   C Conduction of nervous impulses slows down.

   D Body temperature rises.

   E Less water is excreted from the kidneys.

6 Which one of the following crosses from mother to child through the placenta as a result of the mother smoking cigarettes?

   A smoke

   B tar

   C nicotine

   D carbon dioxide

   E sulphur dioxide

7 Which one of the following provides active acquired immunity to the very young child?

   A receiving injections of antigens which cause the body to make antibodies

   B receiving antibodies from mother's milk

   C receiving antibodies from mother via the placenta

   D receiving injections of antigens from the blood of a horse

   E developing antibodies as a result of inherited genes

8 The skin needs to be washed regularly to remove stale

   A hairs.

   B oil.

   C sweat.

   D skin.

   E scales.

9 Which one of the following is the main cause of body odour (BO) when a person has not maintained skin hygiene?

   A the growth and activities of protozoan colonies

   B the failure to keep the hair cut and clean

   C the growth and activities of bacterial colonies

   D the failure to empty the bowels regularly

   E the growth and activities of viral colonies

10 In which of the following habitats are housefly larvae to be found?

   A in the skin of a sheep

   B in a dead rabbit

   C in warm, dry places in the house

   D in stagnant ponds

   E in decaying organic matter

1 The table below shows the numbers of people suffering from infectious diseases in the United Kingdom between 1964 and 1974. Look at the table carefully and then answer the questions below.

	1964	1969	1974
Diphtheria	20	13	3
Food poisoning	6 530	8 599	6 276
Measles	318 912	163 141	118 638
Polio	39	14	8
Tuberculosis (TB)	20 972	14 541	12 456

(a) Which disease affected most people in 1964? . . . . **(1)**

(b) By how many had this disease decreased from 1964 to 1974? (Show all your working)

. . . .

. . . .

. . . . **(4)**

(c) Why had tuberculosis (TB) decreased over the years?

. . . . **(1)**

(d) Which **two** diseases are fairly rare in all years?

1. . . . .

2. . . . . **(2)**

(e) The number of cases of food poisoning was roughly the same in 1964 and 1974. Give a reason why this disease did not decrease very much.

. . . . **(1)**

(f) If you bought a frozen chicken to cook for a family, which two precautions would you take to prevent anyone getting food poisoning? (two lines) **(2)**

(g) What type of organism is responsible for causing TB? (one line) **(1)**

(h) What was the percentage decrease in diphtheria between 1964 and 1969? (Show all your working) (four lines) **(4)**

(i) Which disease is caused by the organism *Salmonella*? (one line) **(1)**

(EMREB)

2 (a) Using the data given in the table below, for a person suffering from influenza, construct a temperature chart on the graph paper provided. **(3)**

Day	1	1	2	2	3	3	4	4	5	5	6	6	7	7	8	8
Time	6 am	6 pm	6 am	6 pm	6 am	6 pm	6 am	6 pm	6 am	6 pm	6 am	6 pm	6 am	6 pm	6 am	6 pm
Oral temperature °C	36.5	37.0	36.5	37.0	37.0	38.0	39.0	40.0	37.5	38.0	37.0	36.5	37.0	37.0	36.5	37.0

(b) Using the information given on your temperature chart, determine the duration of the fever period. **(1)**

(c) List the signs and symptoms of this disease and describe how they may be alleviated. **(3)**

267

(d) Why have immunisation programmes attempting to control this disease been so unsuccessful? **(2)**

(UCLES)

3 (h) Drugs can become addictive. What is meant by the term addictive? (one line) **(1)**

(i) Some drugs like LSD produce hallucinations. What is an hallucination? (one line) **(1)**

(j) Some people who are alcoholics suffer from hallucinations. List four other possible effects of alcoholism. (four lines) **(4)**

(k) (i) Write down three signs you would look for in someone you suspected of glue sniffing. (three lines) **(3)**

(ii) Write down three ways in which glue sniffing is dangerous. (three lines) **(3)**

[Part question] **(EMREB)**

4 (a) 'A vaccine is used to prevent a disease while a serum assists in curing a disease.' Explain the differences between the action of a vaccine and a serum.

A vaccine injected into the bloodstream causes a mild form of the disease and the cells produce antibodies. A serum is the blood containing antibodies against a certain disease which, after preparation, is injected to cure or give temporary immunity. **(7)**

(b) Name **two** diseases against which children are immunised.

(i) *Whooping cough*

(ii) *Diphtheria* **(2)**

(AEB, 1984)

5 (a) Name **two** disease vectors which are commonly found in refuse tips.

(i) *Rats*

(ii) *Houseflies*

(b) Choose **one** of the disease vectors named in (a) and indicate
(i) a disease the vector may carry and
(ii) how the vector may be controlled at a refuse tip.
Name of vector.

Housefly

Disease carried.

Food poisoning

Method of control.

All refuse should be rapidly covered with soil by a bulldozer - prevents access of vector. **(4)**

(AEB, 1985)

6 A pupil carried out the following experiment to show that micro-organisms are present on human hair. The hair was placed on the surface of the sterile nutrient agar contained in the sterile petri dish. The petri dish was sealed, labelled and placed in an incubator for some time. The dish was then observed.

(a) What do you understand by the following words used in this experiment?

(i) sterile *That the agar is free from viable (living) micro-organisms*

(ii) nutrient *This is a food material containing the nutritional requirements of the micro-organisms* **(2)**

(b) At what temperature should the petri dish be incubated? *37°C* **(1)**

(c) Explain why you chose this temperature.

Allows optimum growth of bacteria **(1)**

(d) Which would be the most suitable time for the petri dish to be incubated?
   (i) 4 hours
   (ii) 48 hours
   (iii) 7 days
   (iv) 28 days

   Answer *48 hours* **(1)**

(e) Draw the likely appearance of the agar jelly at the end of the experiment.

(1)

(f) Name two types of micro-organisms that may have grown as a result of this experiment.

(i) *Bacteria*

(ii) *Fungi* **(2)**

(g) Describe the control experiment that the student would also set up.

Sterile nutrient agar in a sterile petri dish that had been open to the air for the same period as the experimental dish with the hair. **(2)**

(h) The pupil now thoroughly washed his hair with a shampoo containing an antiseptic.

(i) What is an antiseptic? *A substance that stops growth of bacteria.* **(1)**

(ii) If the experiment were repeated using a strand of the washed hair, what difference would he have noticed in the appearance of the agar?

There would be fewer colonies of bacteria compared with first experiment. **(1)**

(i) (i) After the experiment had been set up, the teacher never allowed the pupil to open the petri dish. Why?

To prevent further contamination of the agar and to prevent release of any cultured pathogens

(ii) At the end of the experiment, all the dishes were burnt. Why?

To destroy all bacteria that had grown

on the agar, including harmless and harmful types (2)

(LREB)

7 (a) Complete the table for **one** external parasite and **one** internal parasite.

	External parasite	Internal parasite
Name of parasite		
Method of infection		
One method to prevent spread of infection		

(6)

(b) An experiment was set up using various cleaning products to see how effective they were in destroying bacteria. Petri dishes containing nutrient agar were inoculated with bacteria and then the same amount of each of the four products was poured over the agar. The results of the experiment, after four days' incubation, were as shown below.

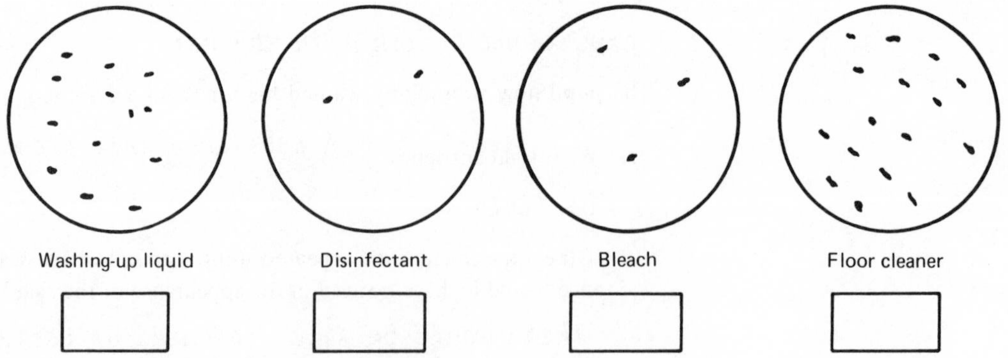

Washing-up liquid	Disinfectant	Bleach	Floor cleaner

(i) Count the number of colonies of bacteria on each dish and enter it in the box beneath the dish. (1)

(ii) Draw a histogram to illustrate the results on the grid provided. Make your own scales. **(4)**

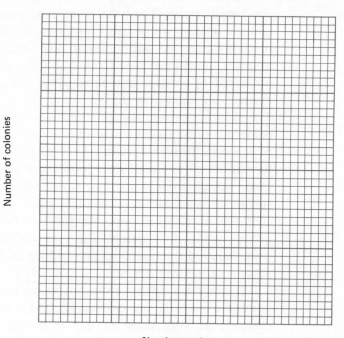

Number of colonies

Cleaning products

(iii) From the results of the experiment, which product would be best for cleaning a lavatory? Give a reason for your answer.

Product ....

Reason ....

.... **(2)**

(iv) From the results of the experiment, which product would be best for cleaning a kitchen chopping board? Give a reason for your answer.

Product ....

Reason ....

.... **(2)**
(WMREB)

Free-response Questions

1 (a) Describe active immunity. **(5)**
   (b) How does passive immunity differ from active immunity? **(4)**
   (c) In what other ways does the body protect itself against infection? **(11)**
(UCLES)

Answer

(a) Antigens are present in the blood as a result of an invasion of the body by a disease-causing organism. They could also be present as a result of an inoculation of live or dead pathogens. These antigens stimulate the white blood cells (lymphocytes) to produce specific antibodies against the antigens (toxins).

The antibodies are produced for some time and are long-lasting in the blood-stream.

(b) Passive immunity develops from an injection of manufactured antibodies into the bloodstream. These antibodies have been obtained from the serum of another animal which had been given the live pathogens. The animal used is often a horse. The antibodies are not long-lasting (unlike those of active immunity), since they are excreted and not replaced. A foetus can be said to be protected by passive immunity, since the antibodies from the blood of the mother can pass across the placenta.

(c) The skin has protective functions against infection. The horny layer of the epidermis prevents the entry of microbes. They can, however, invade the cells of the skin through a wound. Such damage to the skin is sealed rapidly by the formation of a blood clot. Interlacing fibres of fibrin form at the broken blood vessels so that red and white blood corpuscles are trapped. Thus, no further entry of pathogens can take place.

If the microbes do get into the bloodstream, then the white blood corpuscles come into action. The amoeba-like phagocytes ingest the invading organisms and the lymphocytes produce antibodies to neutralise the metabolic poisons of the invaders.

The body has further protective measures, particularly at the possible entry points for bacteria and viruses. The mucous membranes and hairs of the nasal channels are kept moist to trap microbes as we breathe in. The tear glands keep the eyeballs moist and thus wash away dust and bacteria. Wax, which is antiseptic, is produced in the outer ear channels.

If microbes do enter the alimentary canal along with food, the secretion of hydrochloric acid in the stomach is sufficiently strong to kill them. The intake of foods containing vitamin C is important, since this vitamin is vital for the maintenance of healthy mucous membranes and, hence, the destruction of bacteria.

Note

Drawings or diagrams would not help to clarify these answers, and therefore it is best to concentrate on the written prose and ensure precise and clear answers.

2 (a) How does an adult housefly differ from an adult mosquito in the way in which it feeds and in the way in which it spreads disease organisms? **(7)**

(b) Copy this simple diagram (see below) and use it to illustrate the life cycle of the housefly by labelling each box with the name of one of the stages.

Write the word *metamorphosis* alongside the box for the stage at which metamorphosis takes place. **(3)**

(c) Use your knowledge of the life cycle and habits of the housefly to explain how this pest can be controlled. **(10)**

Answer

(a) The housefly has a proboscis with a soft, grid-like structure which can suck in liquid food, whereas the mosquito has mouth-parts which are needle-like and can pierce the skin. The salivary glands of the housefly secrete digestive enzymes onto the food so that it is digested and absorbed in liquid form. On the other hand, the mosquito has saliva which when secreted prevents blood of the host from clotting. This ensures that the blood does not block the tube of the mouth-parts.

The housefly is a carrier of disease-causing organisms on the outside of its body and also in its alimentary canal. The microbes are carried on the bristles of its legs and body, so that when it moves from manure to human food the bacteria are transferred. The mosquito, however, carries the malaria parasites in the saliva and injects them into the blood of the human on which it feeds.

(b)

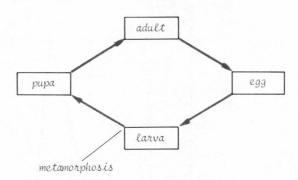

(c) The life history of an insect can be attacked at any one of its stages – egg, larva, pupa or imago (adult). The adult can be destroyed by any of the modern insecticides, especially those that are long-lasting. These are of the spray type, which can saturate the air with piperonyl butoxide around the fly. There are also the organophosphate compounds, which can be incorporated into fly-papers to hang or stand in kitchens. The vapours from these kill off insects which go near them.

The larvae, on hatching from the eggs, move around, feeding on the rotting material in which the eggs are laid. The larvae prefer a moist habitat, so that if the rubbish is kept dry they cannot live. If a compost heap is turned regularly, the internal heat will kill off eggs and larvae that were originally near the surface of the rotting heap. Refuse on which houseflies can breed can be protected by burning or keeping in fly-proof containers, e.g. dustbins with tightly fitting lids. Regular emptying of dustbins by Local Authorities avoids development of larvae that do hatch from eggs.

Larvae when about to pupate seek out a dry habitat. This is not possible when refuse is kept in fly-proof containers. Where there are refuse heaps, it is possible to search for pupae and destroy them.

Note

In part (a) a comparative answer is required, to bring out the differences between feeding method and spread of disease. This must always be done by a point-by-point comparison using such terms as 'whereas', 'on the other hand', and so on (see Chapter 1).

(c) Answers to Objective and Structured Questions

(i) *Multiple-choice Questions*

1. C   2. D   3. D   4. D   5. C   6. C   7. A   8. C   9. C   10. E

(ii) *Structured Questions*

1 (a) Measles   (b) 318 912 − 118 638 = 200 274
   (c) By a process of vaccinating children   (d) Diphtheria and polio
   (e) Through a lack of improvement of hygiene in shops, restaurants and homes.

(f) Defrost completely before cooking; cook at a high temperature for the correct time.   (g) The tuberculosis bacillus (*Mycobacterium*)
(h) $20 - 13 = 7$ $7/20 \times 100 = 35\%$ (i) food poisoning/typhoid

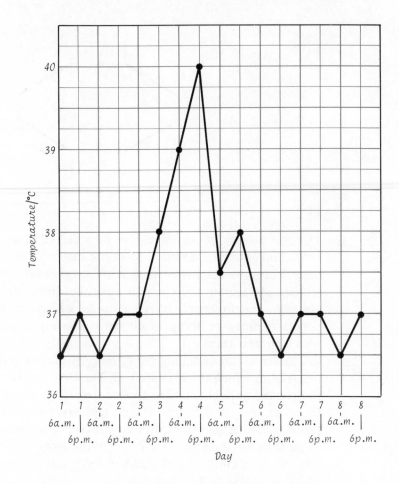

2 (a) See graph.
   (b) 6 a.m. day 3 to 6 a.m. day 6   (c) Fever, sweating, muscular aches, sore throat, headache; take to bed for warmth and take analgesics such as aspirin.
   (d) The virus mutates and changes its form from one epidemic to another. More than one strain of virus can be the cause of infections at any one time.
3 (h) Repeated doses of the drug results in the body becoming 'used to' the presence of the drug, and so the individual becomes ill if the drug is withdrawn.
   (i) The individual will perceive objects not actually present, at first when the eyes are closed but later even when they are open.
   (j) Liver damage; increased pulse rate; insomnia; trembling and sweating
   (k) (i) A strong smell of chemical solvents; stains around the mouth or on the skin; running nose and eyes; nausea and loss of appetite
       (ii) Depressions; nightmares; changes the state of the lungs; death

Questions **4–6** have the answers supplied with the questions. Question **7** has no answers supplied. Try completing the question yourself.

16 Community Health

16.1   Food handling and preservation

Precautions which serve to reduce the incidence of food poisoning are as follows:

1. Storage of food in vermin-proof rooms avoids contamination by faeces of rats and mice.
2. Slaughter and evisceration of animals away from carcass meat or cooked food.
3. Regular disinfection of equipment in slaughterhouses and butchers' shops.
4. Exclusion of any human 'carriers' of disease from slaughterhouses and food shops.
5. Refrigeration of meat and fish; thorough thawing before cooking.
6. Thorough roasting or pressure cooking to achieve high temperatures; many spores of bacteria are resistant to boiling for several hours.
7. Avoiding reheating of food that has been standing in warm conditions.
8. Reducing the handling of food to a minimum and excluding food handlers with septic sores on any part of the body.
9. Education of food handlers in personal hygiene.

16.2   Food processing

(a) Killing Micro-organisms

Micro-organisms can be killed by boiling, roasting or pasteurisation. Cans of food are heated and then stored. Milk will sour within a short time unless pasteurised (kept at $72°C$ for 15 seconds and then cooled rapidly to $10°C$) and refrigerated.

(b) Prevention of Growth and Micro-organisms

1. **Freezing**. Great quantities of food are frozen and moved around the world. Micro-organisms are not killed by freezing; they simply stop growing. Normal household refrigeration at $5°C$ only stops growth for a short period. Deep freezing at below $0°C$ is used for longer periods of storage.
2. **Dehydration**. Water which is essential for growth of micro-organisms is removed. The food is now much lighter and will keep for a long time.

(c) Inhibitors of Growth of Micro-organisms

1. **Acids**. Lactic acid and acetic acid (vinegar) are used for food preservation. They inhibit the growth of bacteria that cannot stand an acid environment.
2. **Salt**. Meat and fish can be kept for long periods without decay in salt, since this stops bacterial growth and enzyme action in the tissues.

3. **Smoke**. Meat products can be cured by smoking which dries the surface and retards the growth of micro-organisms.
4. **Sugar**. Sugar acts as a preservative at high concentrations. Both sugar and salt exert an osmotic action on bacteria. Honey and jam are examples of foods which do not spoil readily.

(d) Radiation

The use of ionising radiation has increased since it has been shown to prevent spoilage and destroy disease-producing organisms.

16.3 Methods of killing bacteria

(a) Physical

1. **Wet heat**. Boiling in water or steam at 100°C is a common method. This does not kill all bacteria unless continued for a long time. Boiling under pressure can raise the temperature to 120°C. All bacteria will be killed at this temperature after 20 minutes.
2. **Dry heat**. Heating to 150°C will kill all bacteria in discarded contaminated dressings. The process destroys the material on which the bacteria are present. Incineration in a furnace is the only method to be used for certain bacteria contaminating dressings.
3. **Pasteurisation**. Named after Louis Pasteur and used particularly for milk. The temperature is raised to 72°C for 15 seconds and then cooled rapidly to 10°C.
4. **Drying**. Bacteria cannot live without moisture. This process is used in the food industry.
5. **Light**. Ultraviolet light kills bacteria. Hence the necessity of letting sunlight into houses and hospitals.

(b) Chemical

1. **Disinfectants** are chemical substances that kill micro-organisms. They are used for washing lavatories, floors, sinks, work-tops and any other place where bacteria could live and grow. Carbolic acid, lysol and cresol are strong disinfectants. Bleach solution, suitably diluted, is best for domestic use.
2. **Antiseptics** are chemical substances that inhibit the growth of bacteria. They are applied to the body in such places as wounds, cuts or burns, to prevent infection. Suitable chemicals are hydrogen peroxide, potassium permanganate and acriflavine. Branded substances such as Dettol contain chloroxylenol and can be used in appropriate dilutions on the skin or in kitchens and bathrooms.
3. **Germicides** are very strong chemical substances that kill bacteria but can also be poisonous to Man, e.g. methanal (formaldehyde), sulphur dioxide and hydrogen cyanide. They can only be used with the greatest care and the operator can only work in protective clothing.
4. **Soaps** are compounds of fatty acids and alkalis. They are used in washing the body to remove natural oils and greases. Bacteria and dirt are trapped in the froth and can then be washed off with clean water. Soaps are also effective for cleaning in the home.
5. **Detergents** are modern substitutes for soap and can be just as effective. They are not used on the skin but are most useful for washing clothes.

16.4 Pollution

(a) Air

1. **Dust**. Small particles produced by industrial processes (e.g. cement works, brickworks).
2. **Smoke**. Industrial furnaces and domestic fires produce about 1000 million kilograms of smoke every year. In certain atmospheric conditions this smoke combines with fog to cause smog.
3. **Sulphur dioxide**. The combustion of coal and oil produces sulphur dioxide. This gas forms sulphurous acid in moist air and, falling in rain, causes corrosion of stone and metal. It also may be responsible for acid rain in Europe, resulting in the death of forests.
4. **Carbon monoxide**, highly poisonous, is produced by petrol engines. It forms high concentrations in rush hour traffic, e.g. in Tokyo policemen directing traffic wear gas masks.
5. **Lead** is produced by petrol engines of cars and lorries. Recent legislation in Europe may result in lowering emissions of this substance during the next ten years.
6. **Smoke from cigarettes** can cause lung cancer in smokers but can also be dangerous to non-smokers.
7. **Nitrogen oxide** is produced by petrol engines and may also be a cause of lung cancer.
8. **Radiation**. Testing of nuclear weapons results in a rise in radiation levels. Nuclear power stations would be dangerous if they went out of control, but the main problem is the disposal of nuclear waste from these stations.

(b) Water

(i) *Fresh water*

1. **Detergents**. Excessive amounts used in washing clothes contain phosphates which pass through sewage works and are discharged into rivers. This water may later be used as drinking-water.
2. **Sewage**. Heavily polluted rivers may result from sewage discharge. Excess bacteria may result in removal of oxygen and death of living organisms (see Fig. 16.1).
3. **Farming. Factory farming** can produce excessive manure, which is often washed away with large amounts of water as slurry. This can pollute rivers and streams.

   Chemical fertilisers used in modern farming are washed into rivers and streams by rainfall. The nitrates in these fertilisers can build up in fresh waters and cause death to living organisms. Furthermore, their presence in drinking-water may reach such high levels that it is dangerous to babies. Water boards must continually monitor the levels of nitrate.
4. **Industrial waste**. Factory waste can contain heavy metals such as lead and mercury together with many dangerous chemicals. These may build up as cumulative poisons in living food chains.

(ii) *Sea-water*

1. **Sewage** is discharged in large amounts from towns bordering the coast. The bacterial content may rise dangerously at certain levels of the tides and be a

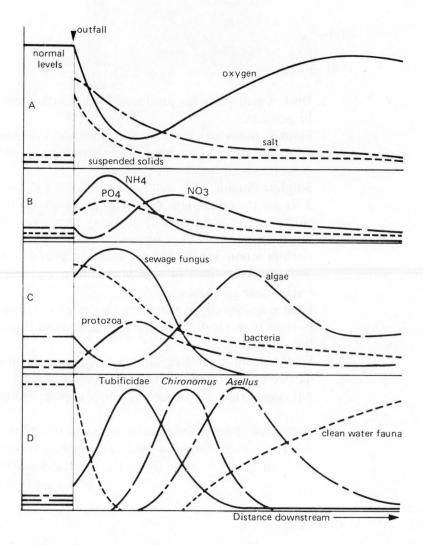

Fig. 16.1 Effect of sewage effluent on a river: A and B, changes in solids and certain chemicals; C, changes in micro-organisms; D, changes in sensitive small animals

danger to sea life and bathers. Inland seas such as the Mediterranean suffer badly from raw sewage and industrial discharge.

2. **Oil**. Leakage from oil tankers and drilling platforms amounts to 5–10 billion kilograms per year. Oil, being a natural product, breaks down in time, but its short-term effect is devastating to sea life.

3. **Metals and chemicals**. Rivers can carry down large quantities of metals and chemicals and they become concentrated as they move up through the food chains.

16.5 Water purification plant

See Fig. 16.3.

Reservoirs store water, which must be treated before communities may draw their drinking-water. The following processes and equipment are involved.

1. **Sedimentation**. Water is pumped into large tanks, and suspended matter sinks to the bottom to be decomposed by saprophytic bacteria.

2. **Coagulation**. Water is pumped into further tanks. Solutions of aluminium sulphate and lime are added, causing fine particles of clay and silt to form larger particles (clumping) and sink on to a settling bed.

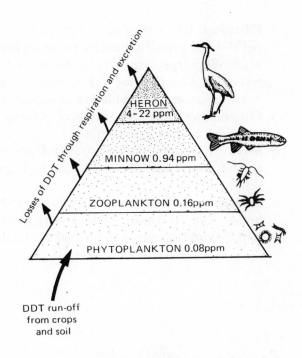

Fig. 16.2 The concentration of DDT up through a fresh-water food chain (units in parts per million)

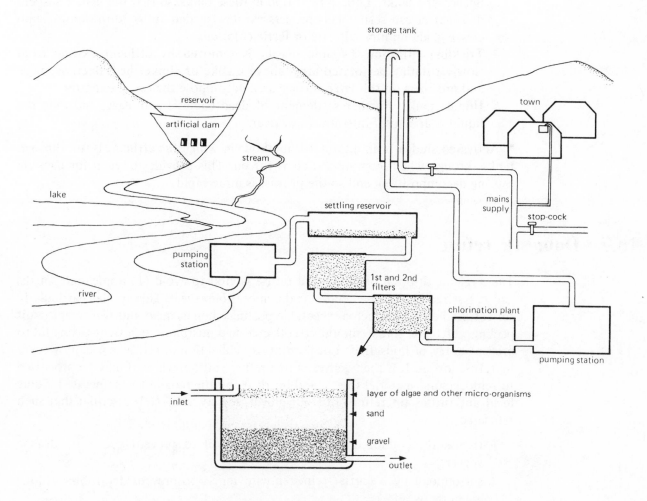

Fig. 16.3 The water supply of a town, with detail of a filter bed

3. **Filtration**. Water runs next through filter beds consisting of layers of sand, pebbles and gravel. Algae and protozoa live on the sand and pebbles, forming a jelly-like layer which helps retain and destroy pathogenic bacteria. The results of filtration are (a) removal of suspended matter; (b) removal of micro-organisms; (c) oxidation of impurities.
4. **Chlorination**. Small quantities of chlorine (0.25 parts per million of water) are added to destroy bacteria.
5. **Service reservoirs**. Clear, filtrated water is stored where there is no further risk of contamination, usually on high ground, in order to flow by gravity into towns.
6. **Storage tanks in buildings**. Usually about 300 dm³ in volume in the roof of a house. Water is drawn off by gravity for washing and cleaning purposes. The tank should be covered to prevent contamination.

16.6 Community sanitation

See Fig. 16.4.

Raw sewage is treated as follows.

1. **Screened**. Passed through metal grids to remove solid matter.
2. **Sedimentation**. Passed into detritus tanks where grit and stones sink to the bottom.
3. **Settlement tanks**. Longer retention in these tanks, so that the lighter suspended matter can settle. This process can be speeded up by addition of chemicals, e.g. aluminium sulphate or ferric chloride.
4. **Trickling filtration**. Organic matter is removed by effluent exiting from slowly rotating perforated arms above a coke or clinker bed. Bacteria, algae and protozoa on this large surface area decompose the organic matter.
5. **Humus tanks**. Further settlement of dead bacteria and algae, and then the liquid is drawn off into stream or river.

Activated sludge is an alternative method, in which the effluent is run through tanks through which compressed air is forced. This provides oxygen for bacteria causing decomposition, and so the process is more rapid.

16.7 Domestic refuse

The hygienic disposal of household refuse is vital to avoid (1) unpleasant smells, and (2) attracting disease vectors (rats, mice, houseflies). This applies particularly to wet kitchen waste such as vegetable peelings, bones, meat and fish scraps, fruit peelings, etc. The refuse should be collected in a dustbin with a tight-fitting lid to prevent entry of houseflies. Local authority collection of refuse is usually weekly by refuse lorries. It is then delivered to a refuse station, where it may be processed to remove solid material (metal and bottles) and the remainder incinerated. Some local authorities tip refuse into old gravel or clay pits. It is essential that such refuse be:

1. frequently bulldozed and covered with soil to prevent access by disease vectors;
2. surrounded by a barrier (chicken-wire fence) to prevent dry rubbish blowing in the wind;
3. if possible, shredded or pulverised mechanically to aid decomposition.

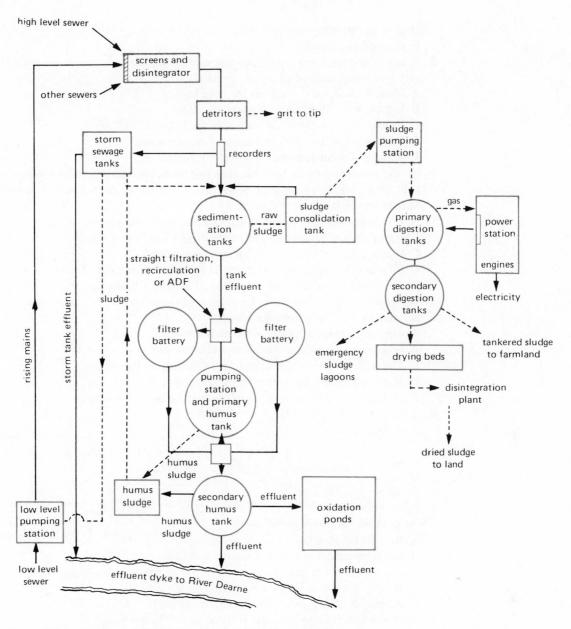

Fig. 16.4 Flow diagram of sewage treatment works

16.8 Questions and answers

(a) Multiple-choice Questions

1 Which one of the following foods would provide a suitable substance for bacterial growth at room temperature?

A smoked fish

B pickled onions

C dried milk

D raisins

E chicken gravy

2 Food, deep frozen at −30°C, has its temperature quickly lowered, so that the bacteria on the food are

A all killed.

B alive and reproducing.

C alive but not growing.

D alive and growing more quickly.

E changed into a viral state.

3 Which of the following results from sewage passing through an aeration tank during the activated sludge method of sewage disposal?

A Waste is dispersed throughout the coke bed.

B Organic waste sinks to the bottom as sludge.

C Bacteria are brought to the surface in the bubbles.

D Oxygen accelerates the breakdown of organic waste.

E Sprinklers circulate and distribute the incoming air.

4 At one time the only method of preserving meat, fish and vegetables was to cover them with large amounts of common table salt. Which one of the following is the effect produced by the salt?

A It forms a covering through which bacteria cannot penetrate.

B It stops houseflies landing and transmitting bacteria to the food.

C It acts as an antiseptic killing bacteria on the food.

D All available water, on which bacteria depend, is drawn out of the food.

E The food heats up through the action of salt and kills the bacteria.

5 Which one of the following is the only certain conclusion that can be drawn if a patient is suffering from *Salmonella* poisoning?

A The food has been contaminated by houseflies.

B Food eaten recently has been raw and uncooked.

C This is the first time the patient has suffered from the disease.

D Infection could have come from either food or water.

E Meat has not been thoroughly refrigerated before consumption.

6 A common pollutant of the air is

A carbon dioxide.

B oxygen.

C smoke.

D nitrogen.

E water vapour.

7 Jam is preserved because of the presence of large amounts of sugar. Which of the following account, for this preservation?

A Bacteria are alive but do not reproduce.

B Fungal spores are not present on the fruit.

C Boiling and the osmotic properties prevent growth of micro-organisms.

D Micro-organisms in the air cannot land on the jam.

E Bacteria and fungi only grow on meat products.

8 The preservation of pickled onions is due to the vinegar being

A acid.

B watery.

C neutral.

D alkaline.

E slightly radioactive.

9 The preservation of canned food is brought about by the removal of

A enzymes.

B water.

C air.

D nitrogen.

E heat.

10 Which of the following temperatures and times is appropriate to the process of pasteurisation (the heat treatment of milk)?

A $70°C$ for 20 minutes

B $100°C$ for 20 seconds

C $100°F$ for 20 minutes

D $70°C$ for 20 seconds

E $37°F$ for 20 seconds

11 One of the developments of modern farming methods has been the increased use of chemical fertilisers for growing crops. One of the disadvantages of this development is that

   A too much food is produced.

   B excess fertiliser may pass into streams, causing increased growth of algae.

   C crops are difficult to harvest.

   D crops used for human food have no flavour.

   E crop plants are weak.   **(SREB)**

12 Joseph Lister developed

   A a cure for tuberculosis.

   B antiseptic surgery.

   C sewage treatment.

   D vaccination against smallpox.

   E the use of ether as an anaesthetic.   **(SREB)**

13 In the treatment of sewage the organic matter from the settlement tanks is made less harmful and unpleasant by the use of

   A filtration.

   B aerobic micro-organisms.

   C sedimentation.

   D methane.

   E anaerobic micro-organisms.   **(SREB)**

14 The main reason why many authorities bury dry refuse (obtained from emptying dustbins) is because it

   A prevents an ugly sight.

   B fills in a big hole in the ground.

   C prevents a smell.

   D prevents rats feeding on it.

   E allows plastics materials to rot quickly.   **(SREB)**

(b) Structured Questions

1 Complete the following table:

Food preservation method	Biological principle involved, including effects on bacteria
Canning beans	
Salting meat	
Deep freezing fish	$-20°C$; stops bacterial growth.
Drying grapes	
Pickling onions	
Refrigerating milk	

            (10)

            (OLE)

2 (a) Sulphur dioxide and carbon monoxide are both pollutants of the atmosphere.

    (i) Name one different source for each of these pollutants. (two lines)   **(2)**

    (ii) State one measure that can be taken to control the amount of sulphur dioxide in the atmosphere. (two lines)   **(1)**

  (b) (i) Name a pollutant of the air from motor vehicles that results from a petrol additive. (one line)   **(1)**

    (ii) What action is being taken in order to reduce this atmospheric pollutant? (two lines)   **(2)**

  (c) Cigarette smoke is a pollutant of air and is a danger to health. State two ways in which action has been taken to attempt to reduce cigarette smoking in this country. (two lines)   **(2)**

283

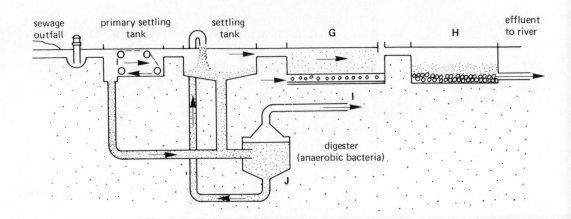

The layout of a sewage treatment plant is shown in the diagram above.

(a) What name is given to this method of sewage treatment? (one line) **(1)**

(b) (i) What form of respiration do the micro-organisms undergo in stage **G**? (one line) **(1)**

    (ii) Name the gas required for this process. (one line) **(1)**

(c) Name the process that occurs at stage **H**. (one line) **(1)**

(d) What gas is given off at stage **I**? (one line) **(1)**

(e) Give one use of the sediment obtained from stage **J**. (one line) **(1)**

(f) This method of processing sewage is much faster than other methods. Name one other method of sewage treatment that is in common use. (one line) **(1)**

(L)

4 (a) By means of a flow diagram outline the treatment of raw sewage from its entry into the sewage works until the treated effluent is released into a river. **(4)**

(b) The figure illustrates the changes that occur in a river downstream from a sewage discharge.

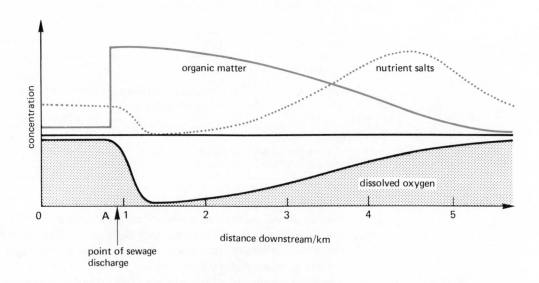

(i) Account for the changes in the concentrations of nutrient salts and organic matter downstream from the point of discharge of sewage at point **A**. **(4)**

(ii) Give reasons for the changes in the concentration of dissolved oxygen downstream from point **A**. **(3)**

(UCLES)

5 (a) The sketch below shows a chain of events which resulted in an outbreak of salmonella (food poisoning) in the people who ate pork sandwiches, but not in the people who ate roast chicken.

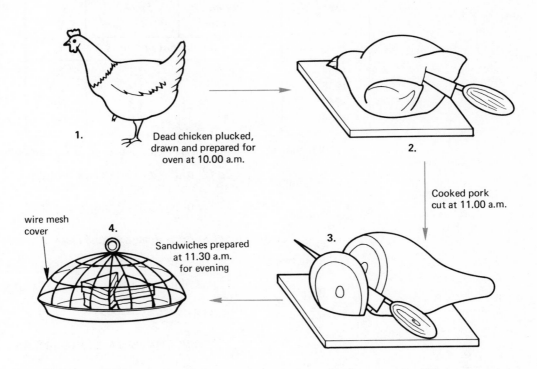

1. Dead chicken plucked, drawn and prepared for oven at 10.00 a.m.

2.

Cooked pork cut at 11.00 a.m.

3.

wire mesh cover

4. Sandwiches prepared at 11.30 a.m. for evening

(i) *Salmonella* causes food poisoning. What kind of organism is *Salmonella*? (one line) **(1)**

(ii) In the events shown above, where did these organisms come from? (one line) **(1)**

(iii) How were these organisms carried to the pork sandwiches? (two lines) **(1)**

(iv) What **three** precautions should have been taken which would have prevented the outbreak of food poisoning? (three lines) **(3)**

(v) Why did the people who ate the roast chicken not suffer from food poisoning? (two lines) **(1)**

(vi) From the sketch, say what precaution was taken to prevent some contamination of the pork sandwiches after they had been made? (two lines) **(1)**

(b) (i) What is an antiseptic? (two lines) **(1)**

(ii) Name the man who first used an antiseptic during an operation? (one line) **(1)**

(iii) Name the substance he used. (one line) **(1)**

(c) If you live near a farm, you may buy milk which is untreated but is from tuberculin tested cows. However, most people buy pasteurised or sterilised milk.

(i) Why is milk pasteurised? (one line) **(1)**

(ii) Describe briefly the process of pasteurisation. (three lines) **(2)**

(iii) If you buy unpasteurised milk from a farm, why does it remain safe to drink for 24 hours? (one line) **(1)**

(iv) Why are cows tuberculin tested? (one line) **(1)**

(d) Although Edward Jenner is regarded as the man who developed vaccination against smallpox in 1796, one form of inoculation was practised in England from 1717. This method involved taking pus from blisters of a smallpox sufferer and transferring it to a healthy person. Often this practice resulted in the healthy person developing a severe form of smallpox.

(i) How did Jenner's method differ from the earlier method? (two lines) **(2)**

(ii) Why was Jenner's method safer? (two lines) **(1)**

(iii) In his original experiment on a boy, how did Jenner show that his method was successful? (two lines) **(1)**

(YREB)

6 (a) The table below shows the amounts of sulphur dioxide, smoke and grit present in the air during the four years 1954, 1958, 1962 and 1966.

Year	Sulphur dioxide	Smoke	Grit
	in millions of tonnes		
1954	5.2	2.1	1.5
1958	6.0	1.7	1.0
1962	6.5	1.3	0.7
1966	6.1	1.0	0.5

(i) What are the main producers of the above mentioned air pollutants?

Sulphur dioxide *Coal and oil burning power stations*

Smoke *Power stations and domestic fires*

Grit *Cement works and domestic fires* **(3)**

(ii) State **two** effects that these air pollutants have on people and buildings.

People 1. *Sulphur dioxide causes lung problems, e.g. bronchitis.*

2. *Smoke prevents sunlight reaching the skin.*

Buildings 1. *Smoke blackens the stonework of buildings.*

2. *Sulphur dioxide forms acid rain and corrodes stonework.* **(2)**

(b) How can noise pollution be reduced in the following situations?

(i) On motorways *Build high embankments with trees to contain the noise of traffic.* **(1)**

(ii) Near airports *Design and build quieter jet engines/double glazing of houses.* **(1)**

(c) (i) When was the Clear Air Act passed in this country?

1956 **(1)**

(ii) What has so far been achieved by the above Act?

The setting up of smokeless zones in cities with the result of more sunlight and less smog. **(3)**

(iii) East Anglia, a primarily rural area, lies to the east of the highly industrialised Midlands (the 'Black Country'). Explain why there is a high level of air pollution in East Anglia.

The prevailing winds are westerly and thus blow pollutants towards East Anglia. **(2)**

(d) For years people have been throwing old car tyres, bottles and cans into village ponds. Although unsightly, the ponds normally continue to support animal and plant life. If, however, a thoughtless car owner dumped a gallon of used oil into the pond, it could well cause total destruction of most visible life in the pond.

(i) Explain how oil is able to kill plant life. *Oil forms a surface film so that gases cannot dissolve and be available to plants. Oil also coats leaves of plants.* **(2)**

(ii) How would oil affect:

1. aquatic insect life? *Oil on the surface prevents access to oxygen of the air.*

2. bird life on the pond? *Oil coats the feathers and the birds are unable to fly.* **(2)**

(YHREB, 1985)

7 (a) What is refuse?

Waste materials from houses and shops. **(1)**

(b) Write down **four** rules for the safe disposal of refuse.

1. *Place in dustbins with well-fitting lids.*

2. *Collect in vehicles that are covered.*

3. *Burn all the organic materials.*

4. *Bury and compress; cover with soil.* **(4)**

(c) Unsafe disposal of refuse could lead to the spread of diseases and cause an epidemic. What is an epidemic?

When a disease outbreak affects many people in an area **(1)**

(d) Name **four** constituents of sewage.

1. *Human faeces*

2. *Human urine*

3. *Household washing-water*

4. *Rainwater* **(4)**

(e) Outline the main stages in the treatment of sewage and briefly explain what happens at each stage.

Crude sewage from the main sewers flows through a grid in which large objects (e.g. sticks) are held back. The sewage then flows through channels, where the grit settles. This is later washed and returned to the land. The next stage is the sedimentation tank, where solid material settles as sludge. This is dried and used as a fertiliser. The effluent passes through a percolating filter; this consists of a concrete tank filled with clinker, coke or small stones. The effluent is sprinkled by rotating arms on to the filter. Micro-organisms feed on organic material (e.g. bacteria) left in the sewage. The effluent finally flows into humus tanks before being pumped into a river. **(10)**

(EMREB)

8 A mystery disease, caused by an unknown microbe, had struck a town. People suffering from the disease were coughing, suffered pains in the chest and a rash appeared on their faces. There were some people who did not contract the disease, although they had been in contact with sufferers. The graph below shows the body temperature of a sufferer after infection. Study the graph and answer the questions which follow.

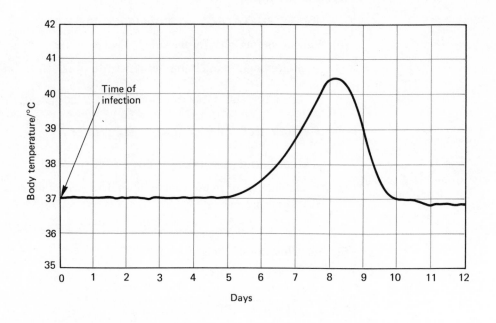

(a) What was the incubation period of the disease?

5 days **(1)**

(b) On which day was the temperature at its highest?

Day 8 **(1)**

(c) What was the highest body temperature reached?

40.5°C **(1)**

(d) What was probably happening to the microbes between

(i) days 5–8? *Dividing rapidly*

(ii) days 8–10? *Dying rapidly* **(2)**

(e) What is the normal body temperature

37.5°C **(1)**

(h) It is unsafe to drink water from natural streams or lakes, because of the presence of pollutants.

(i) What is meant by the term 'pollutant?'

Anything harmful or offensive to animal life

(ii) Name **three** pollutants of water.

1. *Sewage*

2. *Fertiliser chemicals*

3. *Factory chemical effluent* **(4)**

(i) Study the chart below which shows the amount of soap solution needed to form a lather in four water samples. Answer the questions which follow.

Sample	Volume of soap solution/cm³
A	20
B	12
C	8
D	25

(i) Which sample was the hardest water? *D*

(ii) Which sample was the softest water? *C*

(iii) Sample B was temporary hard water. What could you do to make the sample softer?

Boil it and remove temporary hardness. **(3)**

[Part question] **(EMREB)**

9 (a) Micro-organisms which cause disease may enter the body through three different routes. Complete the table to show these routes, an example of a disease which enters by **each** route and **one** natural body defence against each disease.

Route of entry	Disease	Body's natural defence
1	....	....
2	....	....
3	....	....

(9)

(b) (i) Lead and sulphur dioxide are both known to pollute the air. Name **one** different source of each of these pollutants.

Lead ....

Sulphur dioxide .... **(2)**

289

(ii) For **each** of these pollutants, state one measure that can be taken to control the amount present in the air.

Lead controlled by ....

Sulphur dioxide controlled by ....

.... **(2)**

(iii) Name **three** diseases or conditions which may be made worse by cigarette smoking.

1 .... 2 ....

3 .... **(3)**

(iv) State **two** steps that have been taken to try to reduce cigarette smoking in this country.

1 ....

2 .... **(2)**

(c) The histogram below shows the number of outbreaks of food poisoning in Britain in one year.

(i) During which four month period was the total number of outbreaks 17?

.... **(2)**

(ii) In which month was the greatest increase in the number of outbreaks?

.... **(1)**

(iii) In which month did the number of outbreaks decrease by 6?

.... **(1)**

(iv) Give **two** possible reasons for the difference in number of outbreaks in the winter and summer months.

1 ....

2 .... **(2)**

(WMREB)

1 (a) Make a labelled diagram to show, in correct order, the stages by which water from a polluted river is made safe in towns for people to drink. **(10)**

(b) What kind of organism causes cholera? **(1)**

(c) Why is cholera called a water-borne disease? How is cholera spread? **(4)**

(d) What precautions should be taken during a cholera epidemic to prevent more cases developing? **(5)**

(UCLES)

Answer

(a)

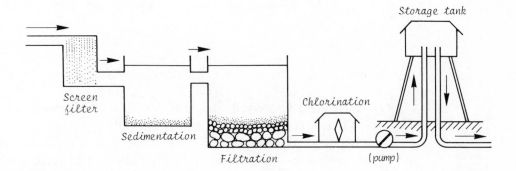

(b) The organism causing the disease cholera is a bacterium of a type called a vibrio.

(c) The pathogen (vibrio) is present in contaminated drinking-water. The pollution of rivers, streams and lakes by untreated sewage containing human faeces may result in reserves of water that are dangerous to drink. This water will infect people with intestinal diseases unless it is boiled or sterilised.

   The disease can also be spread by houseflies moving from infected faeces to food. The bacteria are carried on the legs and body or pass through the gut of the housefly and thus are deposited on the food. The cholera pathogens can again infect the human who consumes the food.

(d) The first task is to check the source of drinking-water. If this is taken from rivers, streams or ponds where there could be faecal contamination, then instructions must be given to boil or chlorinate the water. The community should be told to take great care over the disposal of faeces. If the hygienic methods of lavatory systems and sewage works have been destroyed, then proper latrines must be dug and protected from houseflies.

   The next link in the chain of infection is by way of houseflies settling on food. Instructions must be given that food must be covered or stored in such a way that it cannot be contaminated.

   The individuals that already have the disease must be isolated away from their normal living-quarters. Preferably this should be in a hospital or some temporary accommodation. Great care must be taken over the treatment of soiled clothing and bedding. Stools (faeces) from patients must be properly destroyed. There must also be restrictions on travelling away from the area in which the epidemic is occurring so that carriers or disease-ridden people do not widen the area of infection.

   Finally, if vaccines are available, a widespread programme of vaccination should be commenced to give protection to individuals and to the population as a whole.

Notes

1. The marks allocated to each of these sections indicate the extent of the answer required. This is very clear for the diagram in part (a), where the purification process has five obvious stages. The ten marks will probably be allocated, therefore, to each part of the drawing and its labels or annotations.
2. In part (d) the five marks will be given for clear descriptions of each major action, e.g. treatment of drinking-water.

2 (a) Why is it usually safer to eat freshly cooked meat while it is still hot than after it has been kept for two or three days? (5)
  (b) Explain what is meant when a tin of meat is said to be 'blown'. (5)
  (c) Why is meat a particularly useful kind of food for a growing child? (5)
  (d) Explain why meat can be kept safely for a few days in a refrigerator but should not be left for long at room temperature before being eaten. (5)

(UCLES)

Answer

(a) When meat is cooked, the heat (400°C) is sufficient to kill bacteria. If this meat is still hot, it is safe to eat but, allowed to cool, it could become contaminated. If a fly landed on it while it was still warm, bacteria might be deposited and they would multiply rapidly. The bacteria would digest and incorporate the meat proteins, at the same time producing toxins. The bacteria could be pathogens transferred by the housefly from a dustbin with decaying food. The result to any human that consumed the meat could be food poisoning.

(b) The term 'blown' indicates that the tin has bulged or swollen. The swelling is due to gas present produced by the living processes of the bacteria inside the tin can. The gas produced is carbon dioxide, a by-product of anaerobic respiration.

(c) Meat is mainly protein, and digestion results in the production of amino acids. These are vital molecules for the growth of a child and are an essential part of the diet. The amino acids, minerals and vitamins in meat coming from a mammal (bullock) are most like those found in the body of Man. All three classes of food contribute to growth. Amino acids are required for the production of protein for new cells; minerals and vitamins are required for growth and maintaining the health of the child.

(d) The temperature of the refrigerator is about 5°C and at this temperature bacteria are unable to grow. They are not killed, but for a few days the inhibition of growth ensures that the meat is safe to eat. When the meat is taken from the refrigerator to a room in which the temperature is 15–20°C, the bacteria can resume growth. It is also possible that the meat can be reinfected with pathogenic bacteria capable of producing toxins.

Notes

1. There is no necessity for diagrams in this answer, since straightforward descriptions are required.
2. The question is a composite one, since it requires information on diseases of food poisoning together with the nutritional importance of meat to the growing child.

(c) Answers to Objective and Structured Questions

(i) *Multiple-choice Questions*

1. E   2. A   3. D   4. D   5. D   6. C   7. C   8. A   9. C   10. D   11. B   12. B
13. B   14. D

(ii) *Structured Questions*

1 Canning beans – heat treatment and sealing; kills bacteria
   Salting meat – osmotic action of salt; stops action of bacteria
   Drying grapes – no water present for life; stops action of bacteria
   Pickling onions – acid of vinegar; stops action of bacteria
   Refrigerating milk – low temperatures; stops bacterial growth
2 (a) (i) Power stations and car exhausts   (ii) Washing and filtering smoke emissions from power stations
   (b) (i) Lead   (ii) Lead-free petrol; modification of exhaust systems to remove lead
   (c) Advertising removed from television; cigarette packets must give a health warning against smoking
3 (a) Activated sludge treatment   (b) (i) Aerobic   (ii) Oxygen
   (c) Filtration   (d) Methane gas   (e) Dried and used as fertiliser
   (f) Biological filter treatment
4 (a) grit chamber → settling chamber → trickling filter → humus separator → sludge drying bed
   (b) (i) The build-up of organic matter from the sewage results in the increase of bacteria in the river. This organic matter is broken down by the bacteria, resulting in the production of more nutrient salts.
       (ii) The build-up of organic matter and the increase in the number of bacteria results in a greater use of oxygen. As the water progresses downstream, however, the number of bacteria decrease and less oxygen is removed from the water.
5 (a) (i) Bacillus   (ii) From the chicken   (iii) On the knife/chopping board
       (iv) Wash the knife/use another clean knife; cut the sandwiches and place them in a refrigerator; put pork in a refrigerator and cut sandwiches at the time required.   (v) The chicken was cooked thoroughly and the bacteria were killed.   (vi) A cover was used to protect against anyone touching the sandwiches (it would not protect against houseflies).
   (b) (i) A substance that stops the growth of bacteria   (ii) Lister   (iii) Carbolic acid
   (c) (i) To destroy pathogenic bacteria.   (ii) Milk is heated to 72°C for 15 seconds and then cooled immediately to 10°C.   (iii) It is tuberculin tested and would not go sour within twenty-four hours.   (iv) To ensure that they do not carry *Mycobacterium bovis*, which causes tuberculosis in cattle and Man.
   (d) (i) He did not use pus from smallpox blisters but from related cowpox.
       (ii) This method used by Jenner produces only a small localised pustule on the skin.
       (iii) Six weeks later he inoculated the boy with smallpox and he did not contract the disease.

Questions **6–8** have the answers supplied with the questions. Question **9** has no answers supplied. Try and complete the question yourself.

Index*

*Page numbers in *italics* refer to figures and tables

598.

CENTRAL RESOURCES

R45237A0430

GATEWAY SIXTH
FORM COLLEGE

GATEWAY SIXTH

GATEWAY SIXTH
FORM COLLEGE

Macmillan
Work Out
Series

Work Out

Human Biology

GCSE

The titles
in this
series

For examinations at 16+

Accounting	Human Biology
Biology	Mathematics
Chemistry	Numeracy
Computer Studies	Physics
Economics	Sociology
English	Spanish
French	Statistics
German	

For examinations at 'A' level

Applied Mathematics	Physics
Biology	Pure Mathematics
Chemistry	Statistics
English Literature	

For examinations at college level

Dynamics	Operational Research
Elements of Banking	Engineering Thermodynamics
Mathematics for Economists	

MACMILLAN
WORK OUT
SERIES

Work Out

Human Biology

GCSE

R. Soper

MACMILLAN

© R. Soper 1987

All rights reserved. No reproduction, copy or transmission
of this publication may be made without written permission.

No paragraph of this publication may be reproduced, copied
or transmitted save with written permission or in accordance
with the provisions of the Copyright Act 1956 (as amended).

Any person who does any unauthorised act in relation to
this publication may be liable to criminal prosecution and
civil claims for damages.

First published 1987

Published by
MACMILLAN EDUCATION LTD
Houndmills, Basingstoke, Hampshire RG21 2XS
and London
Companies and representatives
throughout the world

Typeset by TecSet Ltd, Wallington, Surrey
Printed in Great Britain by The Bath Press, Avon

British Library Cataloguing in Publication Data
Soper, R.
Human biology GCSE.—(Work out series)
1. Human biology
I. Title   II. Series
599.9   QP36
ISBN 0-333-42143-4
ISBN 0-333-42144-2 export